Derivatives Algorithms

Volume 1 **Bones**

Derivatives Algorithms

Volume 1 Bones

Tom Hyer

UBS, United Kingdom

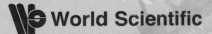

World Scientific

NEW JERSEY · LONDON · SINGAPORE · BEIJING · SHANGHAI · HONG KONG · TAIPEI · CHENNAI

Published by

World Scientific Publishing Co. Pte. Ltd.

5 Toh Tuck Link, Singapore 596224

USA office: 27 Warren Street, Suite 401-402, Hackensack, NJ 07601

UK office: 57 Shelton Street, Covent Garden, London WC2H 9HE

British Library Cataloguing-in-Publication Data
A catalogue record for this book is available from the British Library.

DERIVATIVES ALGORITHMS
Volume 1: Bones

ISBN-13 978-981-4289-80-1
ISBN-10 981-4289-80-9

Printed in Singapore by World Scientific Printers.

Contents

1. Introduction   1

2. Principles   3

    2.1   Our Code   3

        2.1.1   `auto`   4

    2.2   Functional Programming   5

    2.3   Type and State   5

    2.4   Physical Code Structure   6

        2.4.1   Facts   6

    2.5   Platform   7

    2.6   Some Design Patterns   8

        2.6.1   Factory Method   9

        2.6.2   Decorator   9

        2.6.3   Singleton   9

    2.7   Optimization   9

        2.7.1   Calibration   10

        2.7.2   `map`   10

    2.8   Threads   11

3. Types and Interfaces   13

    3.1   The User Base   13

    3.2   A Public Example   14

    3.3   Interface Generation   17

    3.4   Interface Types   18

        3.4.1   Tables and Cells   19

    3.5   Interface Code   21

		3.5.1	Customization Directives	22
3.6		Other Containers		22
3.7		Environment		23
		3.7.1	Exception Messaging	25
		3.7.2	Fast-Path Optimization	27
		3.7.3	Macro Hackery	30
		3.7.4	Repository Access	31
3.8		Enumerated Types		33

4. Vector and Matrix Computations — 37

4.1	Customizing Vectors		37
4.2	Algorithms		38
	4.2.1	Join	40
4.3	Matrices and Square Matrices		41
	4.3.1	Internal Layout	41
	4.3.2	Pasting and Formatting	41
4.4	Matrix Multiplication		43
	4.4.1	Inheritance and Substitutability	44
4.5	Decompositions (Square)		45
4.6	Decompositions (Symmetric)		46
4.7	Decompositions (Sparse)		48
	4.7.1	Tridiagonal Matrices	49
	4.7.2	Band Diagonal Matrices	50
	4.7.3	SLAP Format	51
	4.7.4	The Symmetric Case	55
4.8	Decompositions (Other)		57

5. Persistence and Memory — 59

5.1	Storage		59
5.2	Extraction		62
	5.2.1	Public Types	63
	5.2.2	Example: Linear Interpolant	64
	5.2.3	Reader Registry	66
5.3	Rebuilding		67
	5.3.1	Some Syntactic Sugar	68
5.4	Code Generation		70
5.5	A Display Interface		75
	5.5.1	Storage	75

5.5.2   Extraction 78

5.5.3   Refinements 81

5.6   Auditing 82

5.6.1   Bag 82

5.6.2   Filling Up 83

5.6.3   Audit Types 85

5.7   More on Repositories 87

5.7.1   Unique Objects 87

5.7.2   Naming 89

5.7.3   Matching 90

6.   Testing Framework   91

6.1   Component Tests 91

6.1.1   Physical Structure 93

6.1.2   Reuse 93

6.2   Regression Tests 94

6.2.1   Repository Instrumentation 94

6.3   No Silver Bullet 96

7.   Further Maths   97

7.1   Interpolation 97

7.2   Special Functions 99

7.2.1   The Normal Distribution 99

7.3   Root Solvers 101

7.4   Underdetermined Search 104

7.4.1   Function and Jacobian 106

7.4.2   Weights and Smoothing 108

7.4.3   Monitoring Progress 109

7.5   Quadrature 110

7.5.1   Gaussian Quadrature 111

7.5.2   Adaptive Quadrature 113

7.6   Distributions 113

7.6.1   Implied Vol 114

7.7   Baskets 115

7.7.1   Whole-Basket Moment Matching 116

7.7.2   Taylor Expansion of Projected Vols 118

7.7.3   Midpoint Variance 118

7.8   Random and Quasi-Random Numbers 118

7.8.1   Random Deviate Streams 118

7.8.2   Generator Implementation 119

7.8.3   Transforms 120

7.8.4   Low-Discrepancy Sequences 123

7.8.5   Spectral and Spining Methods 126

7.9   PDE Solvers 127

7.9.1   Cube 128

7.9.2   Coordinate Mapping 129

7.9.3   Coefficient Calculators 131

7.9.4   Forward Induction 133

7.10   American Monte Carlo 134

7.10.1   Recursive Partitioning 134

7.10.2   Biases 136

8.   Schedules   139

8.1   Enumerated Switches 139

8.1.1   Groundwork for Extensibility 141

8.1.2   30E/360 ISDA, ACT/ACT ISMA 144

8.1.3   BUS/252 144

8.1.4   Other Enumerations 146

8.2   Holidays 147

8.2.1   Cities 147

8.2.2   Holiday Sets 148

8.3   Currencies 151

8.3.1   Internals 153

8.4   Increments 154

8.5   Legs 157

8.5.1   Stubs 159

8.5.2   Build from Parameters 160

8.5.3   CDS 162

8.5.4   Inflation Instruments 163

9.   Indices   165

9.1   Naming and Parsing 165

9.1.1   Short Names 167

9.1.2   Nonstandard Indices 167

9.2   Fixings 168

9.2.1   Composites 170

9.3 Sorting and Hashing 171
9.4 Implied Vol 172

10. Pricing Protocols   175

    10.0.1 Which is a Model? 177
10.1 Past and Future 177
10.2 Underlyings 178
10.3 Payments and Streams 178
    10.3.1 Payment Reporting 180
    10.3.2 Commitment to Streams 181
    10.3.3 Destinations 182
10.4 Index Paths 183
    10.4.1 Historical Paths 184
10.5 Defaults and Contingent Payments 185
    10.5.1 Immediate Payments 187
    10.5.2 Viewing Indices 187
10.6 Requests and Promises 188
    10.6.1 Help for Models 190
    10.6.2 Destinations 191
10.7 Bermudans and Barriers 192
10.8 Payouts 194
    10.8.1 Trade State 195
    10.8.2 Values Store 196
10.9 Steps 196
    10.9.1 Valuation and Reevaluation 199
10.10 Use Case Review: PDE 199
10.11 Use Case Review: Monte Carlo and Hedge 201
    10.11.1 Causality 202
10.12 Costs and Benefits 202
10.13 Assembling the Class Hierarchy 203
    10.13.1 Stepper 203
    10.13.2 Asset Values and Tokens 203
    10.13.3 SDE 204
    10.13.4 Model 204
    10.13.5 Trade 205
    10.13.6 Historical Data Access 205
    10.13.7 Assets 207
    10.13.8 Solvers 208

11. Standardized Trades   209

    11.1 Trade Classes 209
    11.2 Cash 211
    11.3 Equity and FX 215
        11.3.1 Equity Forward Payout 215
        11.3.2 Equity Index 217
        11.3.3 Equity Forward Data 218
        11.3.4 FX Option 218
        11.3.5 Forcing Backward Induction 220
    11.4 Legs and Swaps 222
        11.4.1 Putting it Together 225
    11.5 Caps 226
    11.6 Swaps and Swaptions 228
    11.7 Bermudans 230
        11.7.1 Two Views 231
    11.8 Composites 231
        11.8.1 Rescaled Trades 231
        11.8.2 Sums and Collections 233

12. Curves   237

    12.1 Risk 237
    12.2 Libor and Funding 238
    12.3 Build Instruments 239
        12.3.1 Tenor 241
    12.4 Dividend 241
    12.5 Hazard 242

13. Models   245

    13.1 Vasicek-Hull-White 245
        13.1.1 Parametrization 246
        13.1.2 Model Contents 248
    13.2 Interface to Numerical Pricing 249
    13.3 Interface to Valuation Requests 252
        13.3.1 Index Paths 258
        13.3.2 Efficiency 259
        13.3.3 Back to Libor 261
    13.4 Cox-Ingersoll-Ross 262
    13.5 Black-Karasinski 263

13.5.1   Forward Induction PDE Sweep   264
13.6   Single Equity with Local Vol   265
13.6.1   Interpolated Vol   266
13.6.2   Derivation from Implied Vol   266
13.6.3   Model and SDE   267
13.7   A Simple Hybrid Model   268
13.7.1   The Case for Components   270
13.7.2   State Bounds Checks   271

14.   Semianalytic Pricers   273

14.1   A Moment-Matching Pricer   273
14.2   Multimethod Objects   274
14.3   Method Registry   277
14.4   Interaction with Re-evaluator   277
14.5   Interaction with Composites   278
14.6   Pure Pricers   279
14.7   Trade-Dependent Calibration   280
14.7.1   Stabilization   282

15.   Risk   283

15.1   Slides and Bumps   283
15.2   Mutants   284
15.3   Reports   285
15.3.1   Barewords   287
15.4   Portfolios   287
15.5   Tasks   288
15.6   Slide Utilities   289
15.7   Conclusions   290

16.   Additional Code   291

16.1   Add Multiple   291
16.2   ArrayFunctor   291
16.3   Boolean   292
16.4   Composite   292
16.5   Cube   293
16.6   Handle   293
16.7   Matrix   294
16.8   Maybe   297

16.9   PWC (Piecewise Constant) 298

16.10 Vector 299

16.   Acknowledgements and Further Reading   301

Index   303

Chapter 1

Introduction

In June 1994, I spent my last thousand dollars on two suits and a plane ticket to New York. I landed a job in Fixed Income Analytics at Bankers Trust; I had hoped for an offer from Enron, but that was slow to materialize. I quickly realized that I had been hired to produce not mathematics but algorithms.

After Bankers Trust was absorbed into Deutsche Bank, I left for First Union and then for UBS. As I moved into positions of broader responsibility in more successful businesses, the primacy of algorithms became still more pronounced, and I adopted the title of this book as a kind of slogan and began using it on my business cards.

Yet the books available to current and potential practitioners are almost exclusively about mathematics. Many of these books are excellent, but by their nature they can say little about the reality of most quant life. I will try to write about what I really do – about the issues that must fill the attention of a library-building quant.

The result will inevitably be shaped by my own preconceptions and areas of ignorance. I can give any deconstructionists a head start by disclosing my perspective I have been the lead modeller and coder for groups whose main function is to provide a library of analytic functions (models, calibrations thereof, and trade pricing engines) for derivatives pricing. The "desk quants" who study individual trades, and the traders who take final responsibility for the pricing, are my clients. My group's job is to extend the range of what is possible to give a structurer in Hong Kong the power to price a complex hybrid trade, using tools he understands well enough, roughly the way he might understand his car. I am also accustomed to working within a group, and creating protocols that are effectively binding on other coders. The hobbyist pricing complex derivatives in his spare time may find the resulting code too rigid.

Finally, much is omitted here. I do not intend to share insights into my employer's current operations, nor to display the models I consider best. Nor will I betray the reader by publishing castoffs, the failed experiments along the road to working models. Where I cannot write honestly, I will remain silent. Thus the current volume, as its subtitle suggests, demonstrates the creation only of a framework in which powerful models can be used to price very general trades.

Chapter 2

Principles

This is not a book about mathematical methods, nor does it explain any class of models in real detail. (For that purpose, consider the forthcoming *magnum opus* of Leif Andersen and Vladimir Piterbarg, on *Interest Rate Modeling*.) Rather, it is an exposition of the methods needed to describe derivatives and structured financial products in a precise and flexible way, so that both their innate complexity and the complexity of the models for pricing and hedging them them can be controlled. Separation of concerns is a crucial part of this control, and one of our major aims is to demonstrate where this separation may best be accomplished. We also concentrate on the development of reusable components, so we can always understand a given program in terms of large and well-understood atoms; and on a style which encourages clear and concise expression of our intentions.

2.1 Our Code

Most of the code examples presented here will be given in C++. This is partly a result of the author's own experience. In addition, it reflects the expectation that most of the code described will form part of a function library, for potential use by many different applications, and C++ is the library-design language *par excellence*. We will generate a substantial part of this C++ code from higher-level inputs (see "Interface Generation", below); the same techniques will be used to construct interfaces to other languages. We will often describe some concept by showing this *mark-up code* instead of the generated C++ code. The standard template library and Boost libraries will be used freely, and without namespace qualifiers; however, their different stylistic conventions make these contributions easy to spot.

The code examples will inevitably embed our own conventions. The most important of these is that we avoid passing non-`const` references, because they obscure the meaning of user code: does `f(x)` change the value of `x`?[1] This differs sharply from the C++ standard library convention; in particular, we will write `a1.Swap(&a2)`.[2] We also attempt to add information through formatting, using case and underscores to distinguish the names of `Classes_`, `Functions`, `memberData_`, `localVariables` and `function_arguments`.

When a function is nonstandard in some important way or is explicitly not part of an interface, we will put an `X` in front of its name. Think of `X` as standing for "exclusively expert," and signifying that the function should not be called without understanding its implementation. The most common use of this notation is for *ephemeral classes* which hold references whose duration cannot be guaranteed.

The layout of example code is constrained by the 59-column width of the printed page, less than half that of an ordinary modern development screen. I have sometimes compensated for this by using artificially short variable names, which I hope will still convey enough of their purpose to the reader. I have also introduced the macro

```
#define DYN_PTR(n, t, s) t* n = dynamic_cast<t*>(s)
```

solely to reduce line lengths.

2.1.1   `auto`

Typenames in C++ can be quite long, and sometimes depend on implementation details of template libraries. The `auto` keyword, proposed for the future C++0X version of C++, would save effort in several ways. The code examples in this book are written according to the current standard; they nonetheless use `auto`, which has simply been substituted for long type names when this clarifies the code. These substitutions must be undone, which is tedious but never difficult, before most of our examples can be compiled.

---

[1]The C# language addresses this problem admirably: arguments passed by reference must be labeled as such within the calling code.

[2]We will break our own rule at times, when using signaling classes whose whole purpose is to be altered.

2.2 Functional Programming

Our programs will be written in a largely functional style, with most variables and objects being immutable after their construction. This approach is the fundamental strength of "pure" functional languages like Haskell; we make no attempt at purity, but a modicum of discipline provides a great increase in code reusability and maintainability.

The first gain from this style is *referential transparency*, meaning that at each level of the code we need only understand what a function *returns*, not what it *does*. Since errors often arise from obscure conditions encountered during large batch runs in distant lands, this practice simplifies many urgent debugging tasks. This style also leads naturally to code which is less brittle, so that an implementation change is less likely to have unforeseen ill effects.

Within a function body, we will attempt to use `const` values, and to declare them as such. However, we have little compunction about using iterators and such imperative constructs inside functions when they are convenient.

2.3 Type and State

An unfortunately common coding practice is to create an object during one phase of processing, then later to "activate" it in preparation for some anticipated use. For example, a parsed expression from a trade-description language might be activated to prepare for repeated evaluation in different model-generated scenarios. This is invariably a bad practice.

It is easy to see the minor problem: the object before activation is complicated by the empty slots which will be populated during activation, while the object after activation is still carrying around the original data. But this is merely a symptom of a deeper error, which is that the two phases of the object's existence are really two different things, wrongly jammed together in a single object. Another symptom is the absence at compile time of information about the object's state, creating the risk of failure-to-activate.

The solution is now obvious: activation is not a transformation of the initial object, but the creation of a new object. This fits within our functional programming paradigm, and brings the concomitant benefits: referential transparency, more freedom of optimization, compiler assistance in detecting many errors, and more fine-grained separation of concerns. We will see this pattern repeatedly.

2.4   Physical Code Structure

Regardless of language, separation of concerns requires support from the
physical structure of the underlying code. If one piece of code "knows
about" another – *e.g.*, in C++, if it includes the other's header, however
indirectly – then it probably cannot be understood, and certainly cannot
be linked, tested or debugged, without the latter. Additionally, each user
of the former code perforce "knows about" the latter in the same sense,
making changes less local and greatly extending the compilation and link
cycles.

We will always seek to minimize such dependencies. It is common, especially in object-oriented programming, to conceal a class's implementation
from its clients; however, we should go further and to the greatest possible
extent should conceal the *existence* of classes that are not directly needed.
This gives rise to a pattern of separation, at each level of the code: at the
"upper" level are objects that unite several members to perform a coherent
task, using lower-level functions that are unaware of the object's existence.
It also leads to an emphasis on protocols: upper-level objects use abstract
classes embodying lower-level protocols, without any dependency on the
concrete classes which implement those protocols. Since we are writing a
library, not an application, we will never think in terms of a "top" level.
Any attempt to create a top level – *e.g.*, the creation of a static list of
possible implementations of a protocol – should be viewed as a bug.

An important part of minimizing dependencies is to keep information
out of header files when possible. We will rely on *local classes*, implemented
within a local namespace in a single source file, with only a factory function
to create a *new* instance visible in the header. As a rule of thumb, the
aggregate size of source files should be at least six times that of headers in
a passably maintained code base.

2.4.1   *Facts*

Our mindset is toward the development of a functional library, but real-
istically we must expect to export some information to the user as well.
Examples include:

- Lists of known holidays – see Sec. 8.2

- Schedule defaults per currency – see Sec. 8.3
- Bonds deliverable against a bond future

These are part of the computational environment, but it does not really make sense to speak of them as user environment; they are not subject to the user's choice, but are facts about the external world we are modeling.

Such data can be loaded into the library in a plethora of ways, from hardwiring in the code to run-time registry of data sources. The only unifying principle is that we should separate the data acquisition from its use. Thus we do not form a dependency of the data user on the data source, which would reward hardcoding and restrict implementation freedom. Instead, both the data user and an independent data supplier depend on a low-level data storage component.[3]

2.5 Platform

In C++, the contents of a given source file should be described[4] in a header file of the same name, and that header should be *self-contained*: including it should not require the inclusion of some other header. For example, suppose a class `Slide_` is defined in `Slide.h`, and then `SlideEquity.h` defines a class derived from `Slide_`, but that this new header file cannot be compiled without first including `Slide.h`. If we write in `SlideEquity.cpp`

```
#include "Slide.h"
#include "SlideEquity.h"
```

then it will compile without complaint, but a higher-level file which includes `SlideEquity.h` may be confronted with an irritating and unexpected error (which can only be corrected by including `Slide.h` in that file as well).

The usual way to police this is to have each source file include its own header before including other headers, which might hide dependencies. We adopt the policy that each source file must include its own header file as the *second* line of code: the first line of code is reserved for the *platform header* `DA.h`. The platform header thus provides the minimal shared development environment for all code. As such, it has several responsibilities:

(1) Disable otiose warnings (like Microsoft's C4786, symbol greater than 255 characters).

---

[3]Sec. 8.3 shows an example data supplier.
[4]To the extent that they must be available outside that source file.

(2) Optionally, promote other warnings (particularly C4150) to errors.

(3) Forward declare common STL/Boost/Loki classes and interface classes.

(4) Apply **using** declarations and directives to be used everywhere.

(5) Supply some very simple macros, *e.g.* **UNREACHABLE** and **EXPORT**.

(6) Optionally, supply basic math functionality, *e.g.* **Square** and **IsZero**.

Some of the items listed have their own complications: for example, forward declaration is not well supported on all platforms. Our preferred definition of **UNREACHABLE**,

```
──────────────────────── DA.h ────────────────────────
#define UNREACHABLE assert(!"Reached unreachable code")
```

depends on a system-defined **assert** which must be included. It is a good practice to separate host-specific code into its own header files, so **DA.h** might contain:

```
─────────────────────────── DA.h ───────────────────────────
  // ...
  #ifdef _MSC_VER
  #include "DA_MSWindows.h"
  #endif
5 #ifdef SOLARIS
  #include "DA_Solaris.h"
  #endif

  // host-independent code
10 class Dictionary_;
  template<class T_> class Vector_;
  using namespace std;
  // ...
```

Our use of "host" to describe the underlying compiler and hardware, rather than the usual "platform", is deliberate: we wish to emphasize that **DA.h** creates the unique platform on which we will develop.

2.6   Some Design Patterns

A few "design patterns" (in the sense of Gamma *et al.*) do occur repeatedly in our work. These are general programming techniques which, with a little adaptation, can often provide good solutions to a wide variety of design problems. Their use in derivatives pricing has been discussed by

Mark Joshi in some detail,[5] and his book is highly recommended to the interested reader.

2.6.1 *Factory Method*

A factory method is a free function which creates a concrete instance of an abstract class, without disclosing the definition of the concrete class. We will use this to conceal (within a source file) the definition of almost every complicated concrete class.

2.6.2 *Decorator*

A decorator is a class, derived from some abstract interface class, which holds another instance derived from that same interface. It can implement each part of the interface itself, or by forwarding to the member instance, or by forwarding and then "tweaking" the result. The clearest example is a bumped discount curve, which stores the base curve as a member. It computes discount factors by obtaining them from the base curve, then making any necessary adjustment for the bump in rates.

A class cannot safely be decorated if it has *impure* virtual functions; there is too much risk that the decorator will fail to forward the call (*e.g.*, if the base class function's signature changes). For such classes, or if decoration is a common task, we will construct a *null decorator* which simply forwards everything to its internal instance; see Sec. 11.8.1 for an example.

2.6.3 *Singleton*

A singleton is an object which is guaranteed to exist when needed and to be unique within a running application or process. Run-time registries, which are invaluable in reducing compile-time dependencies, must be singletons. We almost always use the Meyers singleton (a static object within an accessor function which returns a reference thereto).

2.7 Optimization

Efficiency of code is an important consideration, which can never be put completely out of mind. However, the sacrifices we are willing to make for

---

[5]sl C++ Design Patterns and Derivatives Pricing, Cambridge University Press, 2004.

efficiency are limited: clarity, functional style, and transparency of intentions are all incommensurably more important in most of our code. Only in known code hotspots will we focus strongly on efficiency.

2.7.1 *Calibration*

It turns out that in practice we pay almost no penalty for this cavalier approach, except in one area: calibration. Because it inevitably entails the creation of high-level objects for each candidate solution, our preference for immutable objects entails more allocation and object initialization than would a more imperative-minded approach. In particular cases like high-frequency trading, we may be forced away from our principles.

It is valuable to distinguish these problem areas as exceptions to be carefully quarantined, rather than imagining that they form a general rule. My own repeated experience has been that "efficiency" is used as a mantra to defend many poor programming practices.

2.7.2 map

A "map", or associative container, is a set of unique keys and an associated value for each; but a `map` is a particular C++ template class, which implements a sorted associative container using a red-black tree. For the most efficient code, we must understand the variety of available containers:

- `map` – insertion, deletion, iteration and lookup are all asymptotically acceptably fast ($O(\lg n)$); but, because each node is heap-allocated, the constant factor is large.
- `hash_map` – from Boost or other libraries. Lookup is substantially faster than a `map`'s, but heap allocation remains a problem. Elements are not sorted, so there is no concept of `lower_bound`.
- `AssocVector` – from Alexandrescu's Loki library. Internally, this is a sorted `vector`, so heap allocation is rarer but insertion can take $O(n)$ time. Iteration is optimal; lookup is as fast as the comparison function will allow.

In this work, we use `map` everywhere by convention. This does not affect the code's correctness, but production code would make different choices for efficiency.

2.8 Threads

Much of the code we will display here is not thread-safe; for example, we use a static repository of in-process objects. In production code, we should guard any access to such an object with a *mutex* – an object supplied by any threading library, which once constructed by one thread will block other threads in its constructor, forcing them to wait until the first thread releases the mutex. In C++, mutexes are inevitably implemented using RAII ("resource acquisition is initialization") – they are locked on construction of a "lock object", and freed in its destructor.

We should obviously avoid sharing non-`const` objects between threads – a corollary is that `mutable` data members, such a cache of results or a vector of workspace, should not be used around threads. To price a path-dependent trade in a multithreaded Monte Carlo, for instance, we must maintain a separate instance of the trade's state for each thread. All our code examples are single-threaded, but should be "future-proof" when (non-const) `static` and `mutable` objects are not directly involved.

This page intentionally left blank

Chapter 3

Types and Interfaces

Some of the code we must produce is nearly independent of the task at hand: the jobs of translating inputs to our own formats, storing and fetching high-level objects, and communicating error messages are all crucial (though less than fascinating). Our first task is to create a supporting system to allow these jobs to be done well, with minimal unnecessary work and boilerplate code.

3.1 The User Base

To create a functioning quant group capable of continuous improvement and re-engineering, we must demarcate what is and is not the work of quants. If pricing tasks are diffused across too large a group, then pricing code will be written in many different locations and will become nearly impossible to track down and change; similarly, a quant group with an over-large mandate will inevitably come to spread pricing code over its entire domain. We avoid this by separating the quant group from its users.

This is a large part of the rationale for our focus on library production: it provides a real-world mechanism for separating pricing and risk tasks (the natural preserve of quants) from database, data transmission, distributed computing and graphical interface tasks. A corollary is that internal types used in the pricing process are the preserve of quants only, and the C++ header files used to build the library are *not* the library interface to its users, which we call the *public* interface.

The concepts we must incorporate into the interface are those which exist outside the process of pricing or risk, or form its top-level inputs: *e.g.* portfolio and risk report. As we create new ways for users to specify these, other concepts will creep into our interface: *e.g.* models, yield curves and

13

even interpolants. But most of our C++ classes will never appear in the public interface, since we often deal in abstractions internal to our own library. The public interface must generally be backward-compatible, but header files away from the public interface may be changed by the quant group without affecting any other group.

3.2   A Public Example

The concepts of the public interface are far easier to understand by example than abstractly. Thus we begin by displaying a public function to query an interpolant. The inputs are a (pre-existing) function object and a vector of x-values at which to interpolate. We describe the function, not in C++, but in a mark-up file:

Interp1_Get.public.if
```
 'Interpolate a value at specified abcissas
 INPUT OBJECT:Interp1 f
 'The interpolant function
 INPUT NUMBER[] x
5'The x-values (abcissas)
 OUTPUT NUMBER[] y
 'The interpolated function values at x-values
```

This is a high-level description of a function, from which C++ code can be generated. The `Interp1` type must have some meaning in C++, which is naturally described in a (non-public) header file:

Interp1.h
```
 #ifndef _INTERP1_H
 #define _INTERP1_H

 #include "Storable.h"
5class Interp1_ : public Storable_
 {
 protected:
     Interp1_(const String_& name)
            : Storable_("Interp1", name) {}
10public:
     virtual double Get(double x) const = 0;
     virtual bool IsInBounds(double) const {return true;}
 };
 #endif
```

Here `Storable_` is our base class for objects visible to library users. In an environment like Excel, such an object will appear to the user as a "handle string" which will be used to look in a `static` singleton repository of `Storable_` objects (see Secs. 3.3 and 5.7).

We wish to separate interface code from library implementation code, so we put it in a different file. Under our preferred naming convention, this source file will be called `__Interp.cpp`:

```
——————————— __Interp.cpp ———————————
#include "Public.h"
#include "Interp1.h"
#include "Functionals.h"
#include "Strict.h"
```

In interface code, we include `Public.h` (which in turn includes `DA.h`) as the first header. This provides definitions of input types like pOPER, and utility functions for translating them. Our preference is to include `Strict.h` as the last header. This allows us to increase and customize warnings *after* including external library code which is not always warning-free.

```
——————————— __Interp.cpp ———————————
   namespace
   {
      struct CheckedInterp_ : unary_function<double,double>
      {
5        const Environment_ mutable* _env;
         const Interp1_& f_;
         CheckedInterp_(_ENV, const Interp1_& f)
            : _env(_env), f_(f) {}
         double operator()(double x) const
10       {
            NOTE(x);
            Require(_env, f_.IsInBounds(x),
                  "X outside interpolation domain");
            return f_.Get(x);
15       }
      };

      void Interp1_Get(_ENV, const Interp1_& f,
         const Vector_<>& x, Vector_<>* y)
20    {
         *y = Apply(CheckedInterp_(_env, f), x);
      }
   } // leave local namespace
```

```
25 //:INSERT\C Interp1_Get.public
```

This much code is handwritten. Next we invoke an *interface gener-ator* which scans the code and finds the instruction (prefaced by //: at the start of a line) to insert code. It reads the nominated file – `Interp1_Get.public.if` – and edits the source file, adding the following immediately after the insertion point:

```
                          __Interp.cpp
   // begin machine-generated code
   extern "C" pOPER mg_xlInterp1_Get
      (const char* xl_f,
      pOPER xl_x)
 5 {
      const char* argName = 0;
      try
      {
         Monitor::CallFromXL("Interp1_Get", "DAMATH.DLL");
10
         Environment::Base_ envBase;
         const Environment_* _env = &envBase;
         argName = "f";
         const Interp1_& f = *Repository::Fetch<Interp1_>
15           (xl_f, false);
         argName = "x";
         Vector_<double> x(XLOPER::ToVector(xl_x, false));

         argName = 0;
20       NOTICE("f", xl_f);
         Vector_<double> y;
         Interp1_Get(_env, f, x, &y);

         Excel::Retval_ retval;
25       retval.Load(y);
         return retval.MakeOper();
      }
      catch (std::exception& e)
      {
30       return XLOPER::Exception(e, argName, "Interp1_Get");
      }
      catch (...)
      {
         return XLOPER::UnknownException("Interp1_Get");
35    }
   }
```

```
XL_REGISTER("Interp1_Get", mg_xlInterp1_Get, "PCP", "f, x")
//-Interp1_get.public -- end machine-generated code
```

Similar code would be added to support a Java, .NET, or other foreign function interface.[1] Even if calling applications were written in C++, we would still expect them to communicate with our code only through this interface – thus defending the separation of quant from non-quant code.

The `LogXLCall` function performs administrative tasks such as checking for library expiration or updating a session usage log; here we deduce the DLL name from the source file path.

The outlines of this code are apparent; this chapter is devoted to explaining the design choices that lead to this particular form.

3.3 Interface Generation

At the public interface of our code library, we need to produce functions for use by applications. Since these applications are not part of our code base – for example, they may be third-party products like Microsoft Excel – the interface should not be presented in terms of our own (C++) types, or of reflections of our types into another language. We will instead have some basic types, and a generic "object" type for internal high-level concepts. Different applications will require different internal representations of the object, such as the handle strings seen by an Excel user.

Like any foreign interface, this requires interface code in our own library combined with additional *mark-up* information for each function. The latter information is not specific to a particular interface; rather, it is a high-level description in our own terms, to be read by our own interface generator.

A further advantage of interface generators is that they can generate multiple outputs. At a minimum, they can produce documentation (*e.g.*, an HTML help page) at the same time as the interface – thus ensuring that the documentation will be always up-to-date, which otherwise requires harsh control over programmers. Also, though only an Excel interface is displayed here, the same information suffices to generate interfaces to Java, .NET, or any other external platform.

---

[1] If we wish to support very many such interfaces, we might interface a single protocol such as SOAP.

The interface generator can be written in many ways.[2] I personally have worked with generators in C++ using hand-written parsers; however, Perl is probably the most appropriate language, especially given the possibility of writing intuitive grammars using the techniques of M.-J. Dominus.[3]

The core of the parser is a table of recognition patterns for each kind of argument, corresponding to Perl classes whose methods will produce each required type of output (HTML help, C++ code, etc.). These methods write code (here, C++ or HTML) for a single argument, like

```
argName = "cpn_basis";
CouponBasis_ cpn_basis(XLOPER::ToString(xl_cpn_basis));
```

This code forms part of a function, all machine-generated, which will catch exceptions and append explanatory information such as the value of the variable `argName`. Once the basic framework is in place, the markup syntax can be extended to support more advanced functionality; see Sec. 3.5.1.

3.4   Interface Types

The fundamental types required should be chosen to be helpful to the user without being too onerous to support. At a minimum, we will support the following types:

- `NUMBER` – a real scalar, presumably double-precision
- `INTEGER`
- `TEXT` – a single text string
- `BOOLEAN` – a true/false value
- `CELL` – a discriminated union of number, text or boolean, like a spreadsheet cell
- `NUMBER[]`, `INTEGER[]`, `TEXT[]`, `CELL[]` – one-dimensional arrays of these types
- `NUMBER[][]`, `CELL[][]` – two-dimensional arrays of these types
- `DICTIONARY` – essentially a `map<String_, String_>`
- `OBJECT` – a `Handle_` to an object of ours, opaque to the user
- `ENUM` – an object of ours, constructible from a string (*e.g.*, call-put flags)

---

[2]I have credible information of an interface generator written as a Microsoft Word macro.

[3]*Higher-Order Perl* by Mark-Jason Dominus, Morgan Kaufmann, 2005.

The latter two entries are not types, but families of types; thus the OBJECT:Interp1 notation in `Interp1_Get.public.if` specifies an object of a particular type. In our code, an ENUM will map to an instance of a class with an enumerable set of valid values; *e.g.*, a coupon basis. If such classes have canonical conversions to and from strings, then they can be constructed with mechanically generated code and passed to our functions; this is detailed in Sec. 8.1.

The following are useful but not indispensable:

- DATE – in this volume we use INTEGER Julian dates
- TIME – a Julian date plus a fraction
- SETTINGS – an object of ours, constructible from a dictionary (*e.g.*, PDE control parameters)
- AUTO – may be a `Handle_` to an existing object, or data from which to construct a new object
- OBJECT[] – a vector of object `Handle_s`

The last three items are again families of types, so we will specify the name of the class being constructed as part of the interface description. Two-dimensional arrays of strings or of integers (TEXT[][] and INTEGER[][]) seem not to be very useful in practice.

3.4.1 *Tables and Cells*

A table is a two-dimensional matrix of cells. We will use the same template `Matrix_` class for all element types – see Sec. 16.7 – so we focus here on defining the CELL.

Boost supplies a **variant** type which implements a discriminated union; a class `boost::empty` (equivalent to the OCaml **unit**) is provided to serve as the type of a variant object with no contents. We choose to explicitly support a Boolean type within the cell, so we have

```
───── DA.h ─────
typedef variant<double, String_, bool, boost::empty> Cell_;
```

The type of a variant is queried using the *visitor pattern* which supplies functionality for each type. This is type-safe, and the compiler will check it for completeness, though the code tends to be bulky. For example,

```
───── TableUtils.h ─────
namespace Cell
{
    struct IsString_ : static_visitor<bool>
```

```
     {
5       bool operator()(const String_&) const {return true;}
        bool operator()(double) const {return false;}
        bool operator()(bool) const {return false;}
        bool operator()(empty) const {return false;}
     };
10   bool IsString(const Cell_& cell)
     { return boost::apply_visitor(IsString_(), cell); }
   }
```

However, the calling code can use utilities like `IsString` and the corresponding `AsString`, so this verbosity does not spread through our code. We can also create a more general type checker (still in `namespace Cell`):

```
────────────────── TableUtils.h ──────────────
   struct TypeCheck_ : static_visitor<bool>
   {
      bool s_, d_, b_, e_;
      TypeCheck_() {s_ = d_ = b_ = e_ = false;}
5
      TypeCheck_ String() const
      { TypeCheck_ ret(*this); ret.s_ = true; return ret; }
      TypeCheck_ Number() const
      { TypeCheck_ ret(*this); ret.d_ = true; return ret; }
10    TypeCheck_ Boolean() const
      { TypeCheck_ ret(*this); ret.b_ = true; return ret; }
      TypeCheck_ Empty() const
      { TypeCheck_ ret(*this); ret.e_ = true; return ret; }

15    bool operator()(const String_& s) const {return s_;}
      bool operator()(double d) const {return d_;}
      bool operator()(bool& s) const {return b_;}
      bool operator()(const empty& s) const {return e_;}

20    bool operator()(const Cell_& cell) const
      { return boost::apply_visitor(*this, cell); }
   };
```

This lets us write checking code like

```
   Require(_env, Matrix::CheckAllElements(correlations, 1,
                 Cell::TypeCheck_().String().Number()),
            "Correlations must be numbers or handle strings");
```

3.5 Interface Code

Why so many types? We could do away with everything except CELL[][]
and construct everything else from it: *e.g.*, an OBJECT is extracted from the
repository using the string value of the sole cell of a 1×1 array. But by
creating more types, we allow more checking to be done in the machine-
generated interface layer, and thus ensure that the checking code will never
be forgotten or neglected, or created by careless copy-and-paste.

In __Interp1.cpp, note the updating of the local argument argName,
which is then used in the catch block to give a more informative error
message. If the translation code (*e.g.*, the call to Repository::Fetch)
fails, the context of the failure will be made apparent.

We take further advantage of this by moving the validity checks for
an argument into the interface description. For example, as part of the
function to construct a cubic-spline interpolant we might write[4]

```
INPUT NUMBER[] x COND{IsMonotonic(x)}    \
   {x-values must be in ascending order}
'The x values at which the function value is specified
```

Our COND construct takes a code snippet as its first argument, and an optional human-readable comment as its second (if the latter is not specified,
then a plausible one is generated from the code snippet). The resulting
C++ code looks like:

```
if (!(IsMonotonic(x)))
   Excel::Throw("x-values must be in ascending order");
```

where the utility Throw throws an exception based on the message given.
The argument name ("x") need not be included in this exception, because
it can be attached later by XLOPER::Exception.

In writing validation code, it is useful to reserve some special character
or string, such as @me, for which the argument name can be substituted
during code generation. This allows constructs like

```
INPUT DATE end_date COND{@me >= start_date}    \
   {@me cannot precede start date}
```

---

[4]We use \ in this example for line continuation, but in real code we would more likely
just use a long single line.

which can be shorter and clearer than the "expanded" version (especially for arguments with multiple constraints).

3.5.1   *Customization Directives*

The interface mark-up for a function may request some customization of the machine-generated code, besides entering an argument list. In particular, a function may be marked as

- A constructor, whose C++ return signature is `Storable_*` (see Sec. 5.1) rather than `void`. The machine-generated code would then handle the returned memory – *e.g.*, for our Excel interface, capture the object in a `Handle_` and store it in the static repository.
- Having direct access to the object repository. The machine-generated code would wrap this access in the input `Environment_`; see Sec. 5.7.

Top-level directives in the `.if` file, such as `CONSTRUCTS Interp1` or `USE_REPOSITORY`, will instruct the interface generator to emit the appropriate code. Further customizations will arise in practice.

3.6   Other Containers

The standard library `string` is adequate for most purposes, except that it is case-sensitive. Quants can forget that, outside the narrow world of C and C++, case is typically ignored. We can use the standard template library "traits" of `std::string` to construct a case-insensitive string; this is a straightforward exercise, but we must be careful to make character comparison as efficient as possible (`toupper` is far too slow for this purpose). The best solution appears to be a static data table, which will be replicated in each source file:

─────────────── *Strings.h* ───────────────

```
namespace
{
   // handwritten lookup table -- specifies ordering
   static const unsigned char CI_ORDER[128] =
         {0, 70, 71, 72, 73, 74, 75, 76, 77, 78,
          79, 80, 81, 82, 83, 84, 85, 86, 87, 88,
// ... (those are non-printing chars at the front)
// traits for case-insensitive string
struct ci_traits : char_traits<char>
{
   typedef char _E;
```

```
      static inline unsigned char SortVal(const _E& _X)
      {
         unsigned char X(_X);
15       return (X & 128) | CI_ORDER[X & 127];
      }
      static bool __cdecl eq(const _E& _X, const _E& _Y)
      {
         return SortVal(_X) == SortVal(_Y);
20    }
      static bool __cdecl lt(const _E& _X, const _E& _Y)
      {
         return SortVal(_X) < SortVal(_Y);
      }
25 };

   typedef basic_string<char, ci_traits> String_;
```

Of course we will choose the ordering so that CI_ORDER['a'] == CI_ORDER['A']. Repetition of this array in (almost) every translation unit increases the size of our DLLs by around 1% – a substantial impact for so little functionality, but not intolerable. We have used the fact that characters beyond 127 are non-printing and their ordering need not have any special properties. We can simplify the logic of ci_traits::eq and lt by using a 256-character lookup table, but this will double the memory use.

We will also customize the STL vector to create our own Vector_: see Sec. 4.1.

3.7   Environment

Despite our preference for pure functions, the problems of derivatives risk management are ineluctably environment-rich. Given a trade; a model representing a distribution of future states of the world of prices; and a history representing past states; there are still other facts which must be specified. For instance:

- What bonds exist which can be delivered against this futures contract?
- Has the Bermudan option to terminate this trade already been exercised?
- Has a given reference credit defaulted according to the terms of this trade?

To address questions like these, we posit a C++ class `Environment_` which will be input to functions which can be affected by it.[5]

The proper design of `Environment_` is itself an interesting illustration of our programming principles. An improperly designed environment would become a chokepoint, bringing together many types of functionality under a common interface which would in turn be seen by any code needing any type of environment (see Sec. 2.4.1). For example, multi-credit trades would be forced to "know about" deliverable bonds.

To avoid this, we will write in C++[6]:

```
──────────────────── Environment.h ────────────────────
namespace Environment
{
   struct Entry_ : noncopyable
   {
      virtual ~Entry_();
   };
}
class Environment_ : noncopyable   // naive implementation
{
public:
   typedef Environment::Entry_ Entry_;
   Vector_<Handle_<Entry_> > vals_;
};
```

A `Handle_` is a shared-pointer-to-`const`; this is the usual use of shared pointers, since a shared pointer to mutable data gives referential transparency to none of its owners. (Later we will see a few specialized uses for shared pointers to non-`const` data, but they are rare; the difficulty of protecting the integrity of input data is a great drawback. Often the only code which can safely use such a pointer is its creator and *de facto* owner, so the pointer is not meaningfully shared.)

The end result is an environment that can be passed to functions which themselves have no idea of the multitudinous forms which the environment may contain, because `Environment_` does not know them itself. This is a precondition of extensible design.

In C++, many functions will require a `const Environment_& env` as part of their declaration. LISP (and Perl 6) users will enjoy the advantage of being able, using function-defining macros, to almost silently receive the

---

[5] More stable facts about the world, such as holiday calendars, are treated differently; see Secs. 2.4.1 and 8.3.

[6] The correct implementation appears in Sec. 3.7.2.

environment. In C++, the closest equivalent is

```
─────────────────── Environment.h ───────────────────
// minimal definition, see below
#define _ENV const Environment_* _env
```

_ENV is always the first rather than the last argument: otherwise its presence would interact with changes to function signatures. The reasons for passing the environment as a pointer, rather than a reference, are explained in Sec. 3.7.3.

3.7.1 *Exception Messaging*

An environment can also serve to capture call stack information which will be displayed if an exception is thrown. We make this possible with an exception class

```
─────────────────── Exceptions.h ───────────────────
  class Exception_ : public std::exception
  {
      Exception_();    // Not implemented
      Vector_<String_> messages_;
5 public:
      Exception_(const String_& complaint);
      void Append(const String_& complaint);
      bool Contains(const String_& complaint) const;
      String_ Display() const;
10 };
```

The **Append** function lets us attach extra information to an exception, which will be visible using **Display**. We could accomplish this with **try** blocks[7]:

```
  try
  {
      // interesting code
  }
5 catch (Exception_& e)
  {
      e.Append("Fitting instrument with maturity "
          + String::FromDate(pi->Maturity()));
      throw;
10 }
```

---

[7]Be sure to use **throw;** rather than **throw e;**, as the latter can lose type and member information from the exception.

This is obviously verbose, but its effects in practice are still worse: what if the information (accessed through the iterator `pi` above) is no longer in scope? To make this approach work, this kind of `try` block must be nested within the code, rather than wrapped around it, leading to repeated `trying` at each level of scope.

To avoid writing such monstrous code, we create a kind of `Environment::Entry_` which will append environment information to exceptions.

```
───────────────────────── Environment.h ─────────────────────────
 class StackInfo_: public Entry_
 {
    const char* name_;
    variant<int, double, const char*, Time_, String_> val_;
 5 public:
    StackInfo_(const char* name, int val);
    StackInfo_(const char* name, double val);
    StackInfo_(const char* name, const char* val);
    StackInfo_(const char* name, const Time_& val);
10  StackInfo_(const char* name, const String_& val);
    void AppendTo(Exception_* e) const
    {
       e->Append(apply_visitor(StackMessage_(name_), val_));
    }
15 };
```

Here we have used a `boost::variant` over the most commonly captured basic types; this can be extended with minimal run-time cost. The `AppendTo` function forwards the relevant stack data to the visitor function object `StackMessage_`. Error checking is then performed by a utility function

```
──────────────────────── Environment.cpp ────────────────────────
 // deprecated implementation
 void Environment::Require(_ENV, bool c, const char* msg)
 {
    if (!c)
 5  {
       Exception_ e(msg);
       const Vector_<Handle_<StackInfo_> > stk
          = Extract<StackInfo_>(env);
       for (auto p = stk.begin(); p != stk.end(); ++p)
10        (*p)->AppendTo(&e);
       throw e;
    }
 }
```

3.7.2 *Fast-Path Optimization*

This code is simple to read, and efficient when an exception is thrown, but our implementation of `Environment_` imposes excessive overhead in the non-exceptional case. In particular, we must create a new environment locally, copy the old contents, and append a `Handle_<StackInfo_>`, both of which require a heap allocation. To prevent this, we need to re-implement `Environment_`:

```
───────── Environment.h ─────────
class Environment_ : noncopyable
{
public:
    virtual ~Environment_();

    typedef Environment::Entry_ Entry_;
    struct IterImp_ : noncopyable
    {
        virtual ~IterImp_();
        virtual bool Valid() const = 0;
        virtual IterImp_* Next() const = 0;
        virtual const Entry_& operator*() const = 0;
    };
    struct Iterator_
    {
        Handle_<IterImp_> imp_;
        Iterator_(IterImp_* orphan) : imp_(orphan) {}
        bool IsValid() const {return imp_.get() != 0;}
        void operator++();
        const Entry_& operator*() const {return **imp_;}
    };

    virtual IterImp_* XBegin() const = 0;
    Iterator_ Begin() const {return Iterator_(XBegin());}
};
```

We provide a function (in `namespace Environment`) to walk through the contents of an `Environment_`:

```
───────── Environment.h ─────────
template<typename F_> void Iterate
    (const Environment_* env, F_& func)
{
    if (env)
    {
        for (auto pe = env->Begin(); pe.IsValid(); ++pe)
            func(*pe);
```

```
        }
    }
```

With this interface it is still easy to implement `Environment::Extract`, and to create a concrete environment `Environment::Base_` which stores a vector of `Entry_`, as in our first-cut implementation. `Base_` will also serve to create empty environments.

A typical accumulator is the one which gathers the stack information for an exception:

```
——————————— Environment.cpp ———————————
struct AccumulateStackInfo_
{
    Exception_* e_;
    AccumulateStackInfo_(Exception_* e) : e_(e) {}
5   void operator()(const Entry_& src)
    {
        if (DYN_PTR(si, const StackInfo_, &src))
            si->AppendTo(e_);
    }
10 };

   void Throw(_ENV, Exception_& e)
   {
       Iterate(_env, AccumulateStackInfo_(&e));
15     throw e;
   }
```

This local code supports the utility functions

```
——————————— Environment.h ———————————
void Require(_ENV, bool c, const char* msg);
void XRequire(_ENV, bool c, const String_ msg);
```

and macros (to avoid unnecessary `String_` construction)

```
——————————— Environment.h ———————————
#define REQUIRE0(c, m) if (c); else throw Exception_(m)
#define REQUIRE(c, m) if (c); else XRequire(_env, false, m)
```

`XRequire` is tagged as expert, not because it is unsafe, but because it can be very inefficient. The argument `msg` must be constructed even if the condition `c` is `true`, and the concomitant string arithmetic can devastate performance.

A common variant of `Extract` is `Extract1`, which returns the "topmost" value of a given type from the environment, or else NULL. It is supported by

```
────────── Environment.h ──────────
template<class T_> struct Extract1_
{
    const T_* found_;
    Extract1_() : found_(0) {}
    void operator()(const Entry_& src)
    {
        found_ = found_ ? found_
            : dynamic_cast<const T_*>(&src);
    }
};
template<class T_> const T_* Extract1(_ENV)
{
    Extract1_<T_> f;
    Iterate(_env, f);
    return f.found_;
}
```

This sometimes unnecessarily walks the whole environment stack, causing an utterly immaterial performance hit.

Also, a derived class can be created which prepends information to an existing environment. We place this in **namespace Environment**; its name reflects the fact that it depends on local references and thus may not be passed out of its local scope.

```
────────── Notice.h ──────────
class XEphemeral_ : public Environment_
{
    const Environment_* parent_;
    const Entry_& val_;

    struct I1_ : IterImp_
    {
        const Environment_* parent_;
        const Entry_& val_;
        I1_(const Environment_* parent, const Entry_& v)
            : parent_(parent), val_(v) {}
        bool Valid() const {return true;}
        IterImp_* Next() const
        { return parent_ ? parent_->XBegin() : 0; }
        const Entry_& operator*() const {return val_;}
    };
public:
    XEphemeral_(const Environment_* p, const Entry_& v)
        : parent_(p), val_(v) {}
```

```
20    IterImp_* XBegin() const
      {
          return new I1_(parent_, val_);
      }
25 };
```

Now creation of a local environment with additional stack information proceeds like

```
StackInfo_ nameInfo("Instrument name", pi->Name());
Environment::XEphemeral_ local(_env, nameInfo);
```

We make this code both simpler and safer by wrapping it in a macro. The
obvious implementation would have us write

```
NOTICE(_env, local, "Instrument name", pi->Name());
```

Using just this macro, it is difficult for a coder to lift an ephemeral
environment off the stack. This should be the only way in which
Environment::XEphemeral_ is used; we can verify this by global text ·
search of the code base.

3.7.3   *Macro Hackery*

The central drawback of this naive macro is the need to specify the name
of the parent and decorated environments (_env and local above), and to
always refer to the most-decorated environment by its ever-changing name.
We can correct this with a version of NOTICE which moves the _env pointer
so that it again refers to the local environment. This is why _ENV defines a
pointer, not a reference, argument. As a first cut, we write[8]

```
——————————————— Notice.h ———————————————
// dangerously incorrect
#define XNOTICE(ee, n, v)    \
    Environment::XEphemeral_ tempEnv_##ee(_env, \
    Environment::StackInfo_(n, v)); _env = &tempEnv_##ee;
5 #define NOTICE2(ee, n, v) XNOTICE(ee, n, v)
  #define NOTICE(n, v) NOTICE2(__COUNTER__, n, v)
```

---

[8]The intermediate NOTICE2 macro is necessary to force __COUNTER__ to be evaluated once,
outside XNOTICE.

However, the end of a local scope could invalidate the temporary environment, leaving `_env` invalid. We must save the parent environment in a locally scoped object which will restore it on departing the scope. Again, we define this object in `namespace Environment`:

```
───────── Notice.h ─────────
struct Save_ : noncopyable
{
    Environment_ const** loc_;
    const Environment_* parent_;
5   Save_(Environment_ const** loc, const Environment_* ch)
        :
    loc_(loc),
    parent_(*loc)
    {
10      *loc = ch;
    }
    ~Save_() {*loc_ = parent_;}
};
```

Now we have a correct implementation of NOTICE.

```
───────── Notice.h ─────────
#define XNOTICE(ee, n, v)     \
    Environment::XEphemeral_ tempEnv_##ee\
        (_env, Environment::StackInfo_(n, v)); \
    Environment::Save_ saveEnv_##ee(&_env, &tempEnv_##ee);
```

Since the name to be logged will often be just the variable name, we provide an additional shorter macro NOTE. We can implement NOTE(x) as NOTICE(#x, x), or we can avoid the need for `__COUNTER__` by using the name of the thing being noticed as part of a (presumably) unique identifier:

```
───────── Notice.h ─────────
#define NOTE(x) XNOTICE(x, #x, x)
```

3.7.4 *Repository Access*

The repository of stored handles is obviously part of the environment: thus our preferred way to access it will be through the `Environment_`. This requires a header file which can include both repository and environment:

```
───────── EnvironmentRepository.h ─────────
#include "Repository.h"
#include "Environment.h"

class ObjectAccess_ : public Environment_::Entry_
```

```
5   {
    public:
        template<class T_> Handle_<T_> Fetch
            (const String_& tag, bool optional) const;
        template<class T_> Vector_<Handle_<T_> > Find
10          (const Pattern_& match) const;
        template<class T_> Vector_<Handle_<T_> > Find
            (const String_& type, const String_* name = 0)
        const;
        template<class T_> String_ Add
15          (const Handle_<T_>& object,
             const RepositoryErase_& erase) const;
        String_ StoredName(const Storable_& object) const;
    };
```

All these functions simply forward to corresponding functions in **namespace Repository**; see Sec. 5.7. They use no member data, but are declared **const** rather than **static** because they should only work if an actual instance of **ObjectAccess_** exists.

We have the option to enforce this as the only form of repository access, but it involves a complicated web of **friends** (for example, **ObjectAccess_** must be given a private constructor, otherwise any function can simply make one). My own experience is that such policing is not necessary: it is enough to provide a simple way to do the right thing.

The most common use of this access is to store a new repository object. The coder, rather than appeal directly to the repository, will call the utility function

─────────── *EnvironmentRepository.h* ───────────
```
    // primitive version without listener
    template<class T_> Handle_<T_> Fetch
        (_ENV, const String_& tag, bool opt = false)
    {
5       NOTICE(tag, "Handle tag");
        auto access = Extract1<ObjectAccess_>(_env);
        Require(_env, opt || access, "No repository access");
        return access
               ? access->Fetch<T_>(tag, opt)
10             ? Handle_<T_>();
    }
```

This is preferred to direct use of `Repository::Fetch`.[9] We will similarly implement utilities to call `Find`, `Add` and `StoredName`. The result is a richer environment which can be provided to functions which request it (see Sec. 3.5.1).

3.8 Enumerated Types

We have mentioned the need for enumerated types, constructible from strings. These can be machine-generated from a high-level description. For example:

```
———————————— OptionType.enum.if ————————————
'Call/put flag
ALTERNATIVE CALL C
ALTERNATIVE PUT P
ALTERNATIVE STRADDLE V C|P
MEMBER double Payout(double spot, double strike) const;
MEMBER OptionType_ Opposite() const;
SWITCHABLE
```

This is sufficient information to define a `class OptionType_` which can be constructed from or converted to a string:

```
———————————— OptionType.h ————————————
class OptionType_
{
public:    // because SWITCHABLE
    enum EOptionType
    {
        CALL,
        PUT,
        STRADDLE,
        _N_VALUES
    };
private:
    EOptionType val_;
public:
    explicit OptionType_(const String_& src);
    const char* String() const;
// Make it SWITCHABLE
    explicit OptionType_(int src);    // for FFI
    OptionType_(EOptionType val);
    EOptionType Value() const {return val_;}
// Idiosyncratic (hand-written) members:
```

[9]At this point, we should consider renaming the direct repository function to `XFetch`.

```
    double Payout(double spot, double strike) const;
    OptionType_ Opposite() const;
};
```

We will also machine-generate the implementations of all but the idiosyncratic members, and of nonmember operators == and !=.

The first text of each **ALTERNATIVE** is the *canonical name* for that value, which will be the output of a conversion to string; additional valid inputs with the same meaning (like "V" to mean **STRADDLE**) may follow. Our parsing of input strings should suppress whitespace; thus

```
OptionType_("c + p").String()
```

evaluates to "STRADDLE" (strings are case-insensitive, as always). We may choose to suppress underscores as well, in which case **ACT360** and **ACT_360** will be considered equivalent.

The **MEMBER** keyword introduces a new member function (to be included in the class definition).[10] Since our entire library is in C++, we can define these members by giving their C++ function signatures; to support multiple languages, we would have to introduce separate keywords like **MEMBER:CPP** for each language.[11]

Because we have made the enumeration **SWITCHABLE**, it can be manipulated directly through its enumerated value. We can consider supporting a public interface based on this value; this will permit coders outside our library to maintain their own option type enumeration, which will be converted to an **int** to pass through our public interface. This cannot work for non-**SWITCHABLE** types, for which we can only accept string inputs; it is probably better to treat all public enumerations in the more general way. We will return to enumerated types in Sec. 8.1, and discuss run-time extension of the set of possible values.

Machine-generated **enum** classes are our preferred method of introducing any control switch. For example, in Sec. 3.7.4 we introduced a type **RepositoryErase_**, which can be defined by creating a mark-up file:

```
───────────────── RepositoryErase.enum.if ─────────────────
'Control what is erased when adding a handle to repository.
ALTERNATIVE NONE
'Erase nothing.
ALTERNATIVE NAME
```

---

[10]We can never introduce new member *data*, which would defeat our purpose.

[11]More precisely, for each language with closed class definitions.

```
 5 │ 'Erase object with the same name.
   │ DEFAULT NAME_NONEMPTY
   │ 'Erase object with the same, nonblank name.
   │ ALTERNATIVE TYPE
   │ 'Erase all objects of the same type.
10 │ SWITCHABLE
```

This mechanism should be used wherever possible; see Sec. 8.1.4.

This page intentionally left blank

Chapter 4

Vector and Matrix Computations

Linear algebra, especially matrix decompositions, and efficient computation with arrays are constant necessities. By carefully structuring the interfaces of these routines, we can write more flexible and safer code.

4.1 Customizing Vectors

The standard library `std::vector` is a fairly good fit for our needs: in particular, its iteration and subscripting are optimal. We would prefer, however, to have somewhat more numerical support.

One important decision is whether to use the same container (or container template) for general-purpose data storage and for processor-intensive arithmetic. By separating the two, we could highly optimize the numerical containers without interfering with the ease of use of storage containers. Our judgement is that the price to be paid in increased code complexity is too steep; we will focus instead on improving `vector`.

We will remove as well as adding functionality; in particular, the two-argument form of `std::vector::resize` is confusing even to experienced programmers.

Our implementation will follow the standard library implementation of the adaptors `stack` and `queue` – these adapt the interface of another object but not its data members, making private inheritance safe. Since the most common element type is `double`, we make that the default, writing simply `Vector_<>` when we need a vector in the mathematical sense.

The `Vector_` class definition (see Sec. 16.10) uses `using` declarations to bring `std::vector` functionality into its interface; functions like the two-argument `resize` are suppressed simply by omitting this declaration. We add member operators `+=` and `-=` (adding/subtracting another vector, or a scalar), and a single `operator*=` to rescale. (Inner product is a nonmember algorithm, not an operator; see Sec. 4.2.)

4.2 Algorithms

The C++ standard library algorithms are deliberately designed for maximum generality, at the expense of making them unpleasantly verbose in many contexts. For example, an in-place `transform` requires the container being operated on to be named thrice – since the name may be nontrivial, such as a nested sub-object or the result of a computation, this leads to cluttered and error-prone code.

Thus we create many container-level algorithms, echoing those defined at iterator level in the STL. When possible, we name these identically to the underlying algorithm except for capitalization; this causes no confusion since the purpose of the algorithm is unchanged.[1]

As an example, here are the implementations of overloaded forms of `Transform`:

```
                         ───── Algorithms.h ─────
   template<class C_, class Op_>
   void Transform(C_* container, Op_ op)
   {
       transform(container->begin(), container->end(),
5            container->begin(), op);
   }
   template<class C_, class CC_, class Op_>
   void Transform(C_* to_modify, const CC_& other, Op_ op)
   {
10     assert(other.size() == to_modify->size());
       transform(to_modify->begin(), to_modify->end(),
            other.begin(), to_modify->begin(), op);
   }
   template<class I_, class O_, class Op_>
15 void Transform(const I_& in, Op_ op, O_* out)
   {
       assert(in.size() == out->size());
       transform(in.begin(), in.end(), out->begin(), op);
   }
20 template<class I1_, class I2_, class O_, class Op_>
   void Transform
       (const I1_& in1, const I2_& in2, Op_ op, O_* out)
```

---

[1] In fact, names that look the same to the human but different to the compiler – allowing disambiguation in some cases – are ideal here.

```
   {
       assert(in1.size() == in2.size());
25     assert(in1.size() == out->size());
       transform(in1.begin(), in1.end(), in2.begin(),
           out->begin(), op);
   }
```

Unless efficiency is at an absolute premium, we prefer a more functional interface which creates and populates a vector in a single step[2]:

─────────────── *Algorithms.h* ───────────────
```
template<class C_, class Op_>
Vector_<typename Op_::result_type> Apply
       (Op_ op, const C_& src)
   {
5      Vector_<Op_::result_type> retval(src.size());
       Transform(src, op, &retval);
       return retval;
   }
   template<class C1_, class C2_, class Op_>
10 Vector_<typename Op_::result_type> Apply
       (Op_ op, const C1_& src1, const C2_& src2)
   {
       assert(src2.size() == src1.size());
       Vector_<Op_::result_type> retval(src1.size());
15     Transform(src1, src2, op, &retval);
       return retval;
   }
```

We extend this notation to other STL algorithms, *e.g.* Copy, LowerBound, BinarySearch, InnerProduct, Accumulate. For the STL unique algorithm, we sort the input as well:

─────────────── *Algorithms.h* ───────────────
```
template<class C_, class LT_>
C_ Unique(const C_& src, const LT_& lt)
   {
       C_ ret(src);
5      sort(ret.begin(), ret.end(), lt);
       ret.erase(unique(ret.begin(), ret.end()), ret.end());
       return ret;
   }
   template<class C_> C_ Unique(const C_& src)
10 {
```

─────────────────────────────

[2]This function is called **map** in most functional languages.

```
    return Unique(src, less<C_::value_type>());
}
```

Since we rarely use `std::list`, we do not bother to specialize the first
version of `Unique`.

Other useful functions, gleaned from those commonly used in pure functional programming, include `Zip`, which generates a vector of pairs from a
pair of vectors; `Filter`, which encapsulates STL's `remove_if`; and `Keys`,
which gets the keys of a `map` (or other associative container).

4.2.1 *Join*

In addition to manipulating vectors, we often want to concisely construct
them from individual elements. This is made possibly by a few template
functions[3]:

Vectors.h

```
    template<class E_> Vector_<E_> Join
        (const E_& e1, const E_& e2)
    {
        Vector_<E_> retval(1, e1);
5       retval.push_back(e2);
        return retval;
    }
    Vector_<String_> Join
        (const char* e1, const char* e2)
10  {
        return Join(String_(e1), String_(e2));
    }
    template<class E_> Vector_<E_> Join
        (const E_& head, Vector_<E_>& tail)
15  {
        Vector_<E_> retval(1, head);
        Append(&retval, tail);
        return retval;
    }
20  template<class E_> Vector_<E_> Join
        (const Vector_<E_>& sofar, const E_& next)
    {
        Vector_<E_> retval(sofar);
        retval.push_back(next);
25      return retval;
    }
```

---

[3]Yes, this is just `consing`.

Thus `Join(Join(1, 2), 3)` constructs a 3-element vector. The template specialization of `Join` for `const char*` is necessary to ensure that it creates a `Vector_<String_>`.

4.3 Matrices and Square Matrices

4.3.1 *Internal Layout*

We choose to use row-major contiguous storage for our `Matrix_` class: this reflects our emphasis on maximal efficiency of access and minimal allocation calls, at the expense of inefficient resizing (since the values must be shuffled around whenever the number of columns changes). We make several requirements of `Matrix_`:

- It must produce ephemeral `Row_` and `Column_` sub-objects which view or manipulate its data;
- These must have `ConstRow_` and `ConstColumn_` relatives for viewing only;
- It must support an ephemeral `SubMatrix_` which sees part of its data, and which also produces `Row_s` and `Column_s`.

As a result, our `Matrix_` implementation is large, and is relegated to the appendices.

4.3.2 *Pasting and Formatting*

This row-major storage scheme means that appending a second matrix by adding extra rows ("append to bottom") is sometimes efficient, while adding extra columns ("append to right") is not. Thus only the former is supported by a special-purpose function:

—————— *Matrix.h* ——————
```
namespace Matrix
{
    template<class E_> void Append
        (Matrix_<E_>* above, const Matrix_<E_>& below);
}
```

For all other combinations, we write a generic function to glue matrices together in a user-specified format. This functionality is bulky enough that we confine it to a source file, exporting it only for the types commonly held in matrices – `double`, `String_` and `Cell_`.

```
                          ──────── MatrixUtils.h ────────
  namespace Matrix
  {
      Matrix_<double> Merge(const String_& format,
          const Vector_<const Matrix_<double>*>& vals);
5     Matrix_<String_> Merge(const String_& format,
          const Vector_<const Matrix_<String_>*>& vals);
      Matrix_<Cell_> Merge(const String_& format,
          const Vector_<const Matrix_<Cell_>*>& vals);
  }
```

These functions take a vector-of-pointers, rather than a vector, as input to avoid unnecessary copying; they do not own or memory manage the pointers. As a rule, when merging only a few matrices, we will just Join their addresses.

The format string defines the operations we might perform. One possible syntax is:

- The numbers 1, 2, ... refer to the elements of the vector in sequence. Note the use of 1-offset.
- The number 0 refers to a 1x1 matrix of an empty Cell_; this can be used to insert spacers.
- The suffix T after a term means the transpose of that term.
- The operators "," and ";" mean side-by-side and top-to-bottom join, respectively (as in Microsoft Excel). These are top-justified and left-justified, respectively.
- The operators "." and ":" are bottom-justified and right-justified variants.
- Parentheses form groups which are then manipulated as units.

This enables formats like 1,0,((2,3,6,7)T:(4,5)) which join several inputs into a highly customized output, without our writing a lot of equally customized code.

The implementation requires a simple parser, and then template functionality to create the output value based on the parsed instructions. This functionality should be implemented within a single source file, to which the Merge functions above are the only interface; thus these functions will be instantiated only once each, avoiding substantial code bloat.

Formatted merges can also be used by Excel::Retval_ (see Sec. 3.2) to let users (or mark-up) control the displayed result of a function with several outputs.

4.4   Matrix Multiplication

With row-major matrix storage, left-multiplication of a vector by a matrix
(*i.e.*, Mv) is extremely efficient.

```
──────────────── MatrixArithmetic.cpp ────────────
template<typename E_> void MultiplyAliasFree
    (const Matrix_<E_>& left, const Vector_<E_>& right,
    Vector_<E_>* result)
{
5    const int n = left.Rows();
    result->resize(n);
    assert(left.Columns() == right.size());
    for (int i = 0; i < n; ++i)
        (*result)[i] = InnerProduct(right, left.Row(i));
10 }
```

The one complication arises from the possibility of aliasing, which is easily
detected because vectors (and our Vector_s) cannot overlap.

```
──────────────── MatrixArithmetic.cpp ────────────
void Matrix::Multiply
    (const Matrix_<double>& left,
    const Vector_<>& right,
    Vector_<>* result)
5 {
    assert(left.Columns() == right.size());
    if (result == &right)      // Support in-place
    {
        Vector_<> temp(right);
        MultiplyAliasFree(left, temp, result);
10    }
    else
        MultiplyAliasFree(left, right, result);
}
```

Our version of `MultiplyAliasFree` is a template which can also be
used for `Matrix_::SubMatrix_` objects. We do not choose to support an
`operator*` for multiplication; in our view, the resulting brevity comes at
too high a cost in clarity. This would change if matrix multiplication were
a constantly recurring task for us; it is not.[4]

For right-multiplication (*i.e.*, $v^T M$) we can either echo the above implementation exchanging columns and rows in the matrix access, or else

---

[4]See Todd Veldhuizen's BLITZ++ library for a sense of the benefits, and costs, of true
dedication to performance.

resort to a more complex implementation that avoids the slower column iterators. This choice is really a matter of taste; experience indicates that right-multiplication by nonsymmetric matrices is not used in any performance hotspots and that the performance difference is small.

Matrix-matrix multiplication raises the same issues again: as well as checking whether the output overwrites the left- or right-hand side, we must also handle the case where all three are the same matrix.[5]

We need one more function: the *weighted inner product*, $v^T A w$.

```
                        MatrixArithmetic.cpp
double Matrix::WeightedInnerProduct
   (const Vector_<>& left,
   const Matrix_<double>& w,
   const Vector_<>& right)
5  {
      assert(left.size() == w.Rows());
      assert(right.size() == w.Columns());
      double retval = 0.0;
      for (int i = 0; i < left.size(); ++i)
10       retval += left[i] * InnerProduct(right, w.Row(i));
      return retval;
   }
```

4.4.1 *Inheritance and Substitutability*

A well-known puzzle of object-oriented programming is to ask, "Is a square a rectangle?" That is, can `Square_` be a subclass derived from `Rectangle_`? Perhaps surprisingly, the answer is often no: a method like `Rectangle_::Resize(double dx, double dy)` cannot be supported by any plausible `Square_`. Similarly, any matrix that supports resizing cannot be a base class for a square matrix.

The solution to this conundrum is in the distinction between `const` and mutable objects: a `const Square_` certainly *is-a* `const Rectangle_`. In C++, we can express this by implementing `Square_` using `Rectangle_ imp_` as the sole member datum, and providing an implicit conversion operator

```
operator const Rectangle_&() const {return imp_;}
```

At this point a `const Square_` can be used anywhere a `const Rectangle_` can. We use exactly this method to ensure that a square matrix, when

---

[5] I am not aware of having written any code which squares a matrix in-place; but I prefer not to find out by blowing it up.

const, is a matrix.[6]

```
 ────────────────────── SquareMatrix.h ──────────────────────
 template<class E_ = double> class SquareMatrix_
 {
     Matrix_<E_> val_;
 public:
 5    SquareMatrix_(int size) : val_(size, size) {}
     operator const Matrix_<E_>&() const {return val_;}
     void Resize(int size) {val_.Resize(size, size);}
     double& operator()(int i, int j) {return val_(i, j);}
     const double& operator()(int i, int j) const
 10          {return val_(i, j);}
     // ...
```

4.5 Decompositions (Square)

We will focus here on the interface, rather than the implementation, of
matrix decompositions. For most purposes we decompose only square matrices, and our interface reflects this.

```
 ────────────────────── Decompositions.h ──────────────────────
 class SquareMatrixDecomposition_ : noncopyable
 {
     virtual void XMultiplyLeft_af(const Vector_<>& x,
          Vector_<>* b) const = 0;
 5    virtual void XMultiplyRight_af(const Vector_<>& x,
          Vector_<>* b) const = 0;
     virtual void XSolveLeft_af(const Vector_<>& b,
          Vector_<>* x) const = 0;
     virtual void XSolveRight_af(const Vector_<>& b,
 10         Vector_<>* x) const = 0;
 public:
     virtual ~SquareMatrixDecomposition_() {}
     virtual int Size() const = 0;      // of the matrix
     // these handle aliasing:
 15  void MultiplyLeft(const Vector_<>& x, Vector_<>* b)
          const;
     void MultiplyRight(const Vector_<>& x, Vector_<>* b)
          const;
     void SolveLeft(const Vector_<>& b, Vector_<>* x)
 20       const;
     void SolveRight(const Vector_<>& b, Vector_<>* x)
```

---

[6]Private inheritance does not work for this purpose, because C++ will find the inheritance – which is inaccessible – before checking for a conversion operator.

```
       const;
};
```

The `MultiplyLeft` and `MultiplyRight` functions compute Ax and xA, respectively, where A is the *original* matrix; `SolveLeft` and `SolveRight` solve $Ax = b$ and $xA = b$. All these are implemented as virtual private expert `_af` functions, reached through a public interface which handles aliasing for them. Behind this interface can live LU, QR and SVD decompositions from *Numerical Recipes*, `LINPACK`, or the proprietary vendor of your choice.

A particular decomposition, necessary for efficient implementation of Markov-chain models, is eigenvalue decomposition of a non-symmetric matrix to permit exponentiation of a matrix. Since not all decompositions can support this, we must extend the interface, using a mixin:

Decompositions.h

```
class ExponentiatesMatrix_
{
public:
    virtual void ExpAT(double t, SquareMatrix_<>* dst)
5       const = 0;
};
```

Thus our `Eigensystem_` will derive from both `SquareMatrixDecomposition_` (or `SymmetricMatricDecomposition_`, below) and `ExponentiatesMatrix_`; a function requiring exponentiation would use `dynamic_cast` to access the extra functionality.

4.6 Decompositions (Symmetric)

The most widely used decompositions, mainly because of their connection to covariance matrices, are those of symmetric matrices. The interface is unsurprising:

Decompositions.h

```
class SymmetricMatrixDecomposition_
    : public SquareMatrixDecomposition_
{
    virtual void XMultiply_af
5       (const Vector_<>& x, Vector_<>* b) const = 0;
    void XMultiplyLeft_af
        (const Vector_<>& x, Vector_<>* b) const
    { XMultiply_af(x, b); }
    void XMultiplyRight_af
```

```
10        (const Vector_<>& x, Vector_<>* b) const
       { XMultiply_af(x, b); }
       virtual void XSolve_af
          (const Vector_<>& b, Vector_<>* x) const = 0;
       void XSolveLeft_af(const Vector_<>& b, Vector_<>* x)
15        const {XSolve_af(b, x);}
       void XSolveRight_af(const Vector_<>& b, Vector_<>* x)
          const {XSolve_af(b, x);}
  public:
       virtual int Rank() const {return Size();}
20     virtual Vector_<>::const_iterator MakeCorrelated
          (Vector_<>::const_iterator iid_begin,
           Vector_<>* correlated)
       const = 0;
       // These handle aliasing:
25     void Multiply(const Vector_<>& x, Vector_<>* b)
          const;
       void Solve(const Vector_<>& b, Vector_<>* x) const;
  };
```

The MakeCorrelated function consumes i.i.d normal deviates from an input vector and populates a vector of correlated multivariate normal deviates whose covariance matrix is described by the decomposition. It returns a pointer to the first unused i.i.d. deviate after the ones it used, which simplifies chaining of such correlators.

Cholesky and eigenvalue decompositions are the most important instances of this type. While the Cholesky decomposition is faster, both to construct and in MakeCorrelated, we will often prefer the eigenvalue decomposition. The latter provides a somewhat meaningful ordering of the input i.i.d. normal deviates in order of importance, which aids variance reduction in Monte Carlo using quasi-random sequences (see Sec. 7.8.5). Often the eigenvalue decomposition can be truncated after a few modes, greatly increasing the speed of simulation with minimal loss of accuracy (naturally, a truncated decomposition can no longer Solve() anything).

The important advantage of the Cholesky decomposition is not its speed, but its stability. In cases where truncation is not feasible and the eigenvalue is not a measure of importance – such as the joint evolution of interest rates, FX and equities in a hybrid model – we will prefer the Cholesky decomposition.

4.7 Decompositions (Sparse)

A large matrix with relatively few nonzero elements is called *sparse*, and there is a substantial literature of mathematical methods for such matrices. The sparse matrices most useful to us are square as well; see Sec. 7.4. A particular task of these matrices is to compute the *Q-form* $JW^{-1}J^T$ used in underdetermined search; see Sec. 7.4. We define them in `namespace Sparse`:

─────────── *Sparse.h* ───────────
```
class Decomposition_ : public SquareMatrixDecomposition_
{
public:
   // form J^T A^{-1} J for given J
5     virtual void QForm(const Matrix_<>& J,
         SquareMatrix_<>* dst) const;
};

class Square_: noncopyable
10 {
public:
      virtual int Size() const = 0;
      virtual bool IsSymmetric() const = 0;

15    virtual void MultiplyLeft
            (const Vector_<>& x, Vector_<>* b) const = 0;
      virtual void MultiplyRight
            (const Vector_<>& x, Vector_<>* b) const = 0;

20    virtual Decomposition_* Decompose() const = 0;

      // element access
      virtual const double& operator()(int i_row, int i_col)
            const = 0;
25    virtual void Set(int i_row, int i_col, double val) = 0;
      virtual void Add(int i_row, int i_col, double val)
      {
         Set(i_row, i_col, val + operator()(i_row, i_col));
      }
30 };
```

The default implementation of `QForm` relies on repeated calls to `SolveLeft`, which is inherited from `Sparse::Decomposition_`; we can override this implementation for symmetric matrices, reducing the number of calls to `SolveLeft` by a factor of two.

A sparse matrix itself does not allow access to the details of its layout; we manipulate it through the `Set` and `Add` functions. By default, `Add` is implemented using `Set`, but we make it virtual to permit optimization. We prefer to return a `const double &` from `operator()`, rather than a `double`, so that incorrect code like `A(i,j)=a` will not compile.

4.7.1 *Tridiagonal Matrices*

The most familiar sparse matrices are tridiagonal, where only elements immediately adjacent to the diagonal are nonzero. A tridiagonal matrix is thus represented by three vectors:

```
————————— Banded.cpp —————————
class Tridiagonal_ : public Square_
{
    Vector_<> diag_, above_, below_;
    double* XAt(int i_row, int i_col);
public:
    Tridiagonal_(int size);
    int Size() const {return diag_.size();}
    bool IsSymmetric() const {return above_ == below_;}
    // ...
```

The accessor `XAt` returns the address of an element in one of the member vectors, or 0 (`NULL`) if the element is too far off-diagonal. In the latter case the `Set` and `Add` functions will throw if the input `val` is nonzero, while `operator()` will return a reference to a local immutable zero value.

This implementation is completely local to `Tridiagonal_`'s source file; in fact, so is `Tridiagonal_` itself. Only a factory function, with a return value of type `Sparse::Square_*`, appears in the header. `Decompose` sends the matrix data to an object of a new type:

```
————————— Banded.cpp —————————
struct TriDecomp_ : Sparse::Decomposition_
{
    Vector_<> diag_, above_, below_;
    Vector_<> betaInvLeft_, betaInvRight_;
    TriDecomp_(const Vector_<>& diag,
            const Vector_<>& above,
            const Vector_<>& below);

    int Size() const {return diag_.size();}
    void XMultiplyLeft_af(const Vector_<>& x, Vector_<>* b)
            const
    {
```

```
        assert(x.size() == Size());
        TriMultiply(x, diag_, above_, below_, b);
15    }
    void XMultiplyRight_af(const Vector_<>& x, Vector_<>* b)
    // ...
};
```

The local free function `TriMultiply` performs either left- or right-multiplication, depending on the order of input arguments. A similar function `TriSolve` implements the decomposition and backsubstitution loop, which may be familiar from *Numerical Recipes*. The difference is that we can take `beta` as an input, having formed it in the constructor; as a result, no temporary vector is needed.[7]

If the computation of β^{-1} fails due to the lack of pivoting, `Tridiagonal_::Decompose` could revert to some more robust method (returning a different subclass of `Sparse::Decomposition_`). We do not implement this refinement.

4.7.2   Band Diagonal Matrices

Band-diagonal matrices generalize tridiagonals, allowing nonzero entries a (small) fixed distance above or below the diagonal. Our class definition and constructor are, as usual, concealed behind a factory function.

```
                           ──────── Banded.cpp ────────
namespace Sparse
{
    Square_* NewBandDiagonal
        (int size, int n_above, int n_below);
5 }
```

Our implementation closely follows that of `Tridiagonal_`, but the `Decompose` function must actually perform a banded *LU* decomposition.[8]

There is one important optimization. If the matrix being decomposed is symmetric, then we can attempt a Cholesky decomposition; this, if it succeeds, gives a more compact representation (taking $\frac{1}{3}$ as much memory) and also supports a more efficient implementation of `QForm`. For, if $A = LL^T$, then $J^T A^{-1} J = (L^{-1} J)^T (L^{-1} J)$; thus we proceed by computing $L^{-1} J$ immediately.

---

[7]Since our convention is to rely mainly on `SolveLeft`, we actually form β_L^{-1} in the constructor, and lazily evaluate β_R^{-1} as needed.

[8]For which we will likely look to *Numerical Recipes*.

```
                        ─── Banded.cpp ───
   void BandedCholesky_::QForm(const Matrix_<>& j,
      SquareMatrix_<>* form) const
   {
      assert(j.Columns() == Size());
 5    vector<Vector_<> > tm(j.Rows());
      // Form L^{-1} J
      for (int ii = 0; ii < j.Rows(); ++ii)
      {
         tm[ii].resize(Size());
10       Copy(j[ii], &tm[ii]);
         BandedLSolve(vals_, tm[ii], &tm[ii]); // in-place
      }
      // Form result
      form->Resize(j.Rows());
15    for (int io = 0; io < j.Rows(); ++io)
      {
         for (int k = 0; k <= io; ++k)
         {
            (*form)(io, k) = (*form)(k, io)
20                = InnerProduct(tm[io], tm[k]);
         }
      }
   }
```

Our use of row-major storage for matrix elements means that iteration along rows is substantially more efficient than along columns. The `vals_` of a `BandedCholesky_` decomposition are oriented so that the inner loop of `BandedLSolve` requires no column iteration; thus it can be made more efficient than the corresponding `BandedLTransposeSolve`. The result is that this version of `QForm`, which requires no computations using L^T, is well over twice as efficient as the default.[9]

4.7.3 SLAP Format

General sparse matrices, whose nonzero elements are not guaranteed to lie near the diagonal, require more sophisticated memory management. The most efficient format for our purposes is a `Vector_<>` of diagonal elements, plus another vector containing for each row (thus our storage is called *row-indexed*) an array of nonzero values and their column locations. This latter is more efficiently implemented using a `Vector_` than a `list`:

---

[9]This is an example of the principles laid out in Sec. 2.3.

```
                         SLAP.cpp
   class SlapMatrix_  : public Sparse::Square_
   {
      Vector_<> diag_;
      Vector_<Vector_<pair<int, double> > > offDiag_;
 5
      void XMultiplyLeft_af
         (const Vector_<>& x, Vector_<>* b) const;
      // ...
      const double& operator()
10       (int i_row, int i_col) const;
      // ...
   };
```

Our algorithms are based on those of the *Sparse Linear Algebra Package*, or *SLAP*.

Methods like `SolveLeft` can no longer be implemented exactly; we must rely on the iterative *conjugate gradient* and, for nonsymmetric matrices, *biconjugate gradient* methods. These are made more efficient by *preconditioning*: rather than solve $Ax = b$ iteratively, we choose a matrix $\tilde{A}$ for which $\tilde{A}\tilde{x} = b$ can be solved efficiently, then solve $\tilde{A}^{-1}Ax = \tilde{A}^{-1}b = \tilde{x}$ iteratively using conjugate gradient.

We provide a mixin which will allow matrices to supply a preconditioner:

```
                          BCG.h
class HasPreconditioner_
{
public:
   virtual void PreconditionerSolveLeft
 5       (const Vector_<>& b, Vector_<>* x) const;
   virtual void PreconditionerSolveRight
         (const Vector_<>& b, Vector_<>* x) const;
};
```

The biconjugate gradient method, which can be found in *Numerical Recipes*, implements `SolveLeft` in terms of `MultiplyLeft`. We implement it as a free function

```
                         BCG.cpp
void BCGSolveLeft
   (const Sparse::Square_& A,
    const Vector_<>& b,
    double tolerance,
 5  int max_iterations,
```

```
Vector_<>* x);
```

We check for a preconditioner using `dynamic_cast`; thus a run-time decision not to precondition (if at compile time we have supplied the inheritance from `HasPreconditioner_`) must be implemented by supplying `Copy(b, x)` or `*x=b` in the body of the `PreconditionerSolve` functions. A useful helper is the ephemeral struct

──────── *BCG.cpp* ────────
```
struct XPrecondition_
{
    const HasPreconditioner_* a_;    // we do not own this
    XPrecondition_(const Sparse::Square_& a)
        : a_(dynamic_cast<const HasPreconditioner_*>(&a)) {}
    void Left(const Vector_<>& b, Vector_<>* x) const
    {
        if (a_)
            a_->PreconditionerSolveLeft(b, x);
        else if (x != &b)
            Copy(b, x);
    }
    // ... similarly for Right
};
```

This lets the solver treat sparse matrices, with and without preconditioning, uniformly.

The simplest way to support `SolveRight` is to call `SolveLeft` with a transposed matrix. Since we do not happen to need transposed matrices in general, we use an ephemeral local struct which holds a reference to A:

──────── *BCG.cpp* ────────
```
struct XSparseTransposed_
    : public Sparse::Square_, public HasPreconditioner_
{
    const Sparse::Square_& a_;
    XPrecondition_ p_;
    XSparseTransposed_(const Sparse::Square_& a)
        : a_(a), p_(a) {}

    int Size() const {return a_.Size();}
    void XMultiplyLeft_af(const Vector_<>& x, Vector_<>* b)
        const { a_.MultiplyRight(x, b); }
    void XSolveLeft_af(const Vector_<>& b, Vector_<>* x)
        const { UNREACHABLE; }
    void PreconditionerSolveLeft(const Vector_<>& x,
        Vector_<>* b) const {p_.Right(x, b);}
```

```
     // ... Right functions are all UNREACHABLE
};
```

Now we can write `BCGSolveRight`, which just calls `BCGSolveLeft` sending `XSparseTransposed_(a)` as the matrix. Since `BCGSolveLeft` does not itself call any `Solve` function, `XSparseTransposed_` should not implement them.

The implementation of a simple function like `MultiplyLeft` gives an idea of how the SLAP layout is used:

─────────────────── *SLAP.cpp* ───────────────────
```
   void SlapMatrix_::XMultiplyLeft_af
      (const Vector_<>& x, Vector_<>* b) const
   {
      assert(x.size() == Size());
 5    b->resize(x.size());
      Transform(x, diag_, multiplies<double>(), b);
      // Now add in off-diagonal elements
      for (int i = 0; i < x.size(); ++i)
      {
10       const auto& r = offDiag_[i];
         double& dst = (*b)[i];
         for (auto e = r.begin(); e != r.end(); ++e)
            dst += e->second * x[e->first];
      }
15 }
```

When a new off-diagonal element is accessed with `Set` or `Add`, we must determine whether it needs to be added to `offDiag_`. A local free function supports this, but it is difficult to handle both the `const` and non-`const` cases with the same code. In the non-`const` case, we can write[10]:

─────────────────── *SLAP.cpp* ───────────────────
```
   double* SlapOffDiag(Vector_<pair<int, double> >& row,
      int i_col, bool create)
   {
      auto ub = row.begin();
 5    while (ub != row.end() && ub->first < i_col)
         ++ub;
      if (ub != row.end() && ub->first == i_col)
         return &ub->second;
      if (!create)
10       return 0;
```

───────────────────────────────

[10]The `while` block here can be replaced by a `for` loop followed by an unexpected semicolon. This is not a wise thing to do.

```
     row.push_back(row.back());
     copy_backward(ub, Previous(row.end()), row.end());
     *ub = make_pair(i_col, 0.0);
     return &ub->second;
15 }
```

This keeps the off-diagonal elements in column order, which is not strictly necessary: an alternative implementation which ignored this ordering might be faster.

This function can support **Set** and **Add** directly. It does not change the **row** if **create** is **false**, so we can use it for read-only access, but must cast away **const**:

SLAP.cpp

```
// unpalatable implementation
const double& SlapMatrix_::operator()
     (int i_row, int i_col) const
{
5    static const double ZERO = 0.0;
     assert(i_row >= 0 && i_row < Size());
     assert(i_col >= 0 && i_col < Size());
     if (i_row == i_col)
        return diag_[i_row];
10   auto& row = const_cast<Vector_<pair<int, double> >&>
           (offDiag_[i_row]);
     double* ret = SlapOffDiag(row, i_col, false);
     return ret ? *ret : ZERO;
}
```

Note that a call to **Set** or **Add** can invalidate the returned reference from **operator()**; this is not a problem in practice.

4.7.4 *The Symmetric Case*

Oddly, *Numerical Recipes* provides code for biconjugate gradient but not the symmetric specialization to conjugate gradient. The latter routine can dispense with half of the preconditioner steps and some vector arithmetic, and is worthwhile to implement as an optimization. The parameter x on input contains a guess at the solution; on output it contains the true solution.[11]

---

[11] The comparison of this routine with the original is also a good demonstration of the expressive power of higher-level programming.

```
                            ─────── BCG.cpp ───────
   void CGSolve
      (const Sparse::Square_& A,
       const Vector_<>& b,
       double tolerance,
 5     int max_iterations,
       Vector_<>* x)
   {

       const int n = A.Size();
       assert(b.size() == n && x->size() == n);
10     assert(!IsZero(tolerance) && max_iterations > 0);

       double tNorm = tolerance * sqrt(InnerProduct(b, b));
       XPrecondition_ precondition(A);
       Vector_<> r(n), z(n), p(n);
15     A.MultiplyLeft(*x, &r);
       Transform(b, r, minus<double>(), &r);   // r = b - Ax
       double bkden;
       for (int ii = 0; ii < max_iterations; ++ii)
       {
20         precondition.Left(r, &z);
           const double bknum = InnerProduct(z, r);
           const double t = ii > 0 ? bknum / bkden : 0.0;
           Transform(z, p, AddMultiple(t), &p);
           bkden = bknum;
25         A.MultiplyLeft(p, &z);
           const double ak = bknum / InnerProduct(z, p);
           Transform(x, p, AddMultiple(ak));
           Transform(&r, z, AddMultiple(-ak));
           if (InnerProduct(r, r) <= tNorm)
30             return;
       }
       throw Exception_("Exhausted iterations in CGSolve");
   }
```

One useful preconditioner for symmetric sparse matrices is the *incomplete Cholesky decomposition*. Here we attempt a Cholesky decomposition, with the constraint that the decomposition may have nonzero values only where the source matrix itself has a nonzero value. This decomposition is, as its name suggests, incomplete; $LL^T \neq A$. An alternative preconditioner is the tridiagonal part of A, ignoring all other nonzero entries; this is less similar to A but easy to form and rapid in application. These three possibilities (incomplete Cholesky, tridiagonal part, or no preconditioner) must be tested in realistic cases to see which provides the best performance; there is no clear winner in all scenarios.

Band diagonal matrices should be favored when possible due to the speed and precision of the banded LU and Cholesky decompositions. The generic form is used only in some advanced applications, such as simultaneous underdetermined calibration of multiple coupled surfaces.

4.8 Decompositions (Other)

A diagonal or lower-triangular matrix is already "decomposed" enough for our purposes. It is useful to provide utility functions which implement `SquareMatrixDecomposition_` for such matrices, which often arise in Jacobians for chaining of risk sensitivities. Naturally, the header file does not display the concrete class:

```
——————— DecompositionsMisc.h ———————
SymmetricMatrixDecomposition_* DiagonalAsDecomposition
   (const Vector_<>& diag);
SquareMatrixDecomposition_* LowerTriangularAsDecomposition
   (const SquareMatrix_<>& src);
```

Singular value decomposition, which supports linear fitting, is another useful tool. We do not display it here, but there is one failing of the common implementation which we should point out. The function

```
xInv = x > xMin ? 1.0 / x : 0.0;
```

which is used in place of the simple $1/x$ when inversion of the singular values is called for, is gratuitously discontinuous. Continuity is necessary for small input changes to lead reliably to small value changes, and desirable in many other ways. Thus we use instead

```
xInv = x > xMin ? 1.0 / x : x / Square(xMin);
```

or even

```
xInv = x / (Square(x) + Square(xMin));
```

where xMin is, as usual, a user-supplied fraction of the maximal singular value.

This page intentionally left blank

Chapter 5

Persistence and Memory

To support both debugging of complex tasks and distributed computing, we will need to provide object persistence – the ability to save an object or transfer it to a different process, generally by writing it to a file or database.

Some languages offer built-in support for this, or have a sufficiently simple object model that a general-purpose persistence library can be built. In the Python language, for example, any object is essentially a nested hash table, and the "pickle" module provides object-to-file conversions for most objects. Erlang goes even further, with built-in concurrency and little distinction between persisted and in-process data. However, partly because of the constraints required to support this universality, these languages are not really suitable for our purposes.

More general-purpose languages often come with libraries which support persistence of objects built within that library: examples include CL-STORE for LISP and MFC for C++.

However, there are significant advantages to writing our own persistence methods, which as far as I am aware have not been replicated in any widely available library. Foremost among these is the automatic maintenance of documentation, as mentioned above. Also, serialization is nearly conceptually identical to inspection, and we will support both without duplicative code.

5.1 Storage

We will define a *storable* base class for objects which can be saved and recovered ("persisted" in the regrettable lexicon of quants). Writing functions which save to some concrete storage mechanism – e.g., XML – is a common mistake. Obviously, this practice binds the code to a single type of storage and wastes the opportunity to create new implementations behind an abstract interface. In practice, it also leads to the proliferation of different persistence idioms; these might all be valid, but they act in concert to destroy the uniformity of stored data and of persistence code.

Thus `Storable_` should know only about an abstraction of archiving:

```
────────────────── Storable.h ──────────────────
     class Storable_;
     namespace Archive
     {    // forward declarations
        class Dst_;
 5      void Write(Dst_&, const Storable_&);
        void XWrite(Dst_&, const Handle_<Storable_>&);
        class Data_;
     }


10   class Storable_
     {
        virtual void Save(Archive::Dst_& dst) const = 0;
        friend void Archive::Write
           (Archive::Dst_&, const Storable_&);
15      friend void Archive::XWrite
           (Archive::Dst_&, const Handle_<Storable_>&);
     public:
        const String_ name_;
        const String_ type_;
20      Storable_(const char* type, const String_& name);
        virtual ~Storable_();
     };
```

The `type` is supplied as a C-style string, rather than a `String_`, simply to check that the two inputs to the constructor are not inadvertently swapped. `Archive::Dst_` is the interface through which an object writes itself.

```
────────────────── Storable.h ──────────────────
     class Dst_
     {
        // support store-by-reference
        virtual bool StoreRef(const Storable_& object) = 0;
 5      virtual bool StoreRef
           (const Handle_<Storable_>& object) = 0;
        friend void Archive::Write
           (Archive::Dst_&, const Storable_&);
        friend void Archive::XWrite
10         (Archive::Dst_&, const Handle_<Storable_>&);
     public:
```

```
     virtual ~Dst_();
     // create fundamental types by direct assignment
     virtual void operator=(bool val) = 0;
15   virtual void operator=(double val) = 0;
     virtual void operator=(const String_& val) = 0;
     virtual void operator=(const Time_& val) = 0;
     virtual void operator=
        (const Vector_<String_>& val) = 0;
20   virtual void operator=(const Vector_<>& val) = 0;
     virtual void operator=
        (const Matrix_<Cell_>& val) = 0;
     virtual void operator=(const Dictionary_& val) = 0;
     // create composite types; these auto-vivify
25   virtual Dst_& Child(const String_& name) = 0;
     virtual Dst_& Element(int i) = 0;
     // support polymorphism
     virtual void SetType(const String_& type) = 0;
     // sugar for WriteToArchive
30   template<class T_> void operator=
        (const Handle_<T_>& object)
     { Archive::Write(*this, object); }
};
```

The operator= functions set the archive to contain one concrete datum; such an archive needs no further type information, and should never be given any. The StoreRef functions are needed to support storage by reference, so that shared sub-objects can be written only once and afterwards a unique identifying tag can be substituted.[1] These functions return true if such a reference can be found; then the write is complete. More likely they will return false, meaning that the object is not written, and we must now write it. This demand – write if and only if StoreRef returns false – simplifies the writer's implementation, because StoreRef can do any necessary writing when it succeeds, and can create a tag for future reference when it fails. The two overloaded functions exist because Handle s can be manipulated – e.g., stored in a repository – in ways that bare objects cannot.

This gives us the tools to write to some abstract archive any of the fundamental types in the interface above, or any object composed of members which can themselves be archived. Thus we are now in a position to define the friends of Storable_ and Archive::Dst_:

---

[1] The tag is seen only by the archive, not by the object.

```
─────────────── Storable.h ───────────────
   namespace Archive
   {
      void Write
         (Archive::Dst_& dst,
 5        const Storable_& object);
      void XWrite
         (Archive::Dst_& dst,
          const Handle_<Storable_>& object);
      template<class T_> inline void Write
10       (Archive::Dst_& dst,
          const Handle_<T_>& obj)
      {
         XWrite(dst, HandleCast<Storable_>(obj));
      }
15 }
```

The `Write` functions ensure that `StoreRef` is called correctly. The `Save` member in a derived class should never call `StoreRef`; thus the latter is `private` in `Dst_`.

One subtlety is that the `type_` of a `Storable_` object will generally not be the type sent to `Dst_::SetType`. The former is mainly a hint or comment to users, while the latter is specific to a particular concrete class (and, in a mature system, to a version of the object's serialization; see Sec. 5.4).

5.2 Extraction

We can reconstitute such types from an archive through a query interface which is the mirror image of the writing interface above (also in `namespace Archive`):

```
─────────────── Storable.h ───────────────
   class Src_
   {
   public:
      virtual ~Src_();
 5    // query fundamental types
      virtual Maybe_<bool> AsBoolean() const = 0;
      virtual Maybe_<double> AsNumber() const = 0;
      virtual Maybe_<String_> AsString() const = 0;
      virtual Maybe_<Time_> AsTime() const = 0;
10    virtual Maybe_<Vector_<> > AsVector() const = 0;
```

```
     virtual Maybe_<Vector_<String_> > AsStringVector()
        const = 0;
     virtual Maybe_<Matrix_<Cell_> > AsTable() const = 0;
     virtual Maybe_<Dictionary_> AsDictionary() const = 0;
15   // query composite types
     virtual Handle_<Archive::Data_>& Contents() const = 0;
     virtual String_ Type() const = 0;
     virtual map<String_, Handle_<Src_> > Children()
        const = 0;
20   virtual vector<Handle_<Src_> > Elements() const = 0;
     // notify of unexpected children
     virtual void Unexpected(const String_& cn) const = 0;
   };

25 class Data_
   {
   public:
     virtual ~Data_();
     virtual void Save(Archive::Dst_& dst) const = 0;
30 };
```

The extraction of fundamental types is straightforward: the only guarantee we need is that an archive leaf element created with a particular operator= in `Archive::Dst_` must return the same value from the corresponding query function. In practice other queries will likely succeed as well: for example, a number will be stored in an XML file as text, and thus could be read out as a `String_` unless we clutter the XML with extra type information. It is better to keep the archive simpler, since the "accidental" success of a query does no harm in practice.

But we have sneaked a new class, `Archive::Data_`, into this interface. In principle we could construct a `Storable_` directly from the archive; but this would require all `Storable_` objects to implement a suitable constructor interface with a specific argument ordering. It would also constrain the internal layout of each `Storable_`, which would have to hold all its members in the canonical way expected by the archive. The role of `Data_` is to hold the necessary data and simultaneously obey all these constraints.

5.2.1 *Public Types*

Thus, beside the inheritance hierarchy of `Storable_`, there is a parallel hierarchy of `Data_`. To complete the read from archive, we need to construct the former from the latter.

The base `Data_` class cannot accomplish this; its only role is to provide a homogenous method for data storage. Derived classes will define a built type and a build method; but again, there is a complication.

Recall that we have the option to store objects by reference, to eliminate the need for redundant storage of reused objects. The `Src_::Contents` function supports this, by allowing both query for a previously read `Data_` and notification of a newly read one. But now two or more handles to `Data_`, stored disparately in a larger structure, must both be built to `Storable_` objects. If the build processes are uncoupled, not only will we duplicate the computational effort of the build, but the resulting objects will no longer be shared: they will be different objects with all their member data duplicated. Thus the reconstituted object will take up more memory, and if it is saved again, it will create a larger archive with duplicated data. To avoid this, the build function must take a *build record* which maps previously built data to the final object, and can be used to see whether subobjects have already been built.

While the base `Data_` class lacks a `Build` method, each concrete `Data_` object has one returning an appropriate built type. Several concrete `Data_` types might display the same interface, because they build different variants of a common `Storable_` object. For example, we might build several different concrete one-dimensional interpolants, but handle them all through a single base class for storage purposes. These base classes are the building blocks of higher-level objects. For example:

```
──────────────────────── Interp1.h ────────────────────────
struct Data_Interp1_ : Archive::Data_
{
    typedef Handle_<Interp1_> built_t;
    virtual built_t Build(Archive::Built_* built)
5       const = 0;
};
```

This defines an abstract `Data_` type which is known to be usable to construct an `Interp1_`. The set of all such derived abstract types is the *public type set*, so named because the classes it describes are visible, and often useful, to our external users.

5.2.2 *Example: Linear Interpolant*

We would next write

```
──────────────── InterpLinear.cpp ────────────────
   struct Data_Interp1Linear_v1_ : Data_Interp1_
   {
      Vector_<> x;
      Vector_<> f;
 5    built_t Build(Archive::Built_* built) const;
      void Save(Archive::Dst_& dst) const
      {
         Save_Interp1Linear_v1(dst, x, f);
      }
10 };
   Data_Interp1_::built_t Data_Interp1Linear_v1_::Build
      (Archive::Built_*) const
   {
      return Handle_<Interp1_>(Interp::NewLinear1(x, f));
15 }
```

The reader code, which populates data based on an archive source, mirrors the **Save** method above:

```
──────────────── InterpLinear.cpp ────────────────
   Handle_<Archive::Data_> ArchiveExtract_Interp1Linear_v1
      (_ENV, const Archive::Src_& src)
   {
      NOTICE("Extracting", "Interp1Linear, version 1");
 5    auto_ptr<Data_Interp1Linear_v1_> data
            (new Data_Interp1Linear_v1_);
      map<String_, Handle_<Archive::Src_> > cMap
            = src.Children();
      auto psrc_x = cMap.find("x");
10    Require(_env, psrc_x != cMap.end(), "No element 'x'");
      Require(_env, !psrc_x->second.Empty(),
            "Element 'x' is NULL");
      const Archive::Src_& src_x = *psrc_x->second;
      Require(_env, src_x.AsVector().Known(),
15          "Element 'x' is not a numeric vector");
      data->x = src_x.AsVector().Value();
      cMap.erase(psrc_x);

      auto psrc_f = cMap.find("f");
20    Require(_env, psrc_f != cMap.end(), "No element 'f'");
      Require(_env, !psrc_f->second.Empty(),
            "Element 'f' is NULL");
      const Archive::Src_& src_f = *psrc_f->second;
      Require(_env, src_f.AsVector().Known(),
25          "Element 'f' is not a numeric vector");
```

```
      data->f = src_f.AsVector().Value();
      cMap.erase(psrc_f);
      for (auto pu = cMap.begin(); pu != cMap.end(); ++pu)
      {
30        src.Unexpected(pu->first);
      }
      Require(_env, data->x.size() == data->f.size() ,
            "x and f must have the same size");

35    Handle_<Archive::Data_> object(data.release());
      return src.Contents() = object;
}
```

This code is awkward due to the verbose error checking (see Sec. 5.4, where we show how to machine-generate it), but accomplishes the job of extracting an object which may have been created by our own **Save** method, or by some other process outside our control.

5.2.3   *Reader Registry*

Now that we can extract **Data_** of one particular type, we must prepare for objects whose exact type is unknown. But these concrete **Data_** types are scattered across the code, and isolated in individual objects. A generic reader must be able to find them, which of course is the purpose of the **Type** tag of an archived object. The concrete readers are simply function pointers:

—————————————— *Storable.h* ——————————————
```
namespace Archive
{
    typedef Handle_<Data_>(*reader_t)(const Src_&);
}
```

We provide a central singleton registry (a **map<String_, reader_t>**), and export the functions **RegisterReader** and **UnregisterReader**. The former is called at DLL load time; the latter is called at DLL unload, if we choose to support this.

Now, given an archived object, a centralized function can look at its **Type** and dispatch to the appropriate reader:

—————————————— *Storable.cpp* ——————————————
```
Handle_<Archive::Data_> Archive::Read
    (_ENV, const Archive::Src_& src)
{
```

```
     if (Boolean_<Handle_<Data_> > old = src.Contents())
5        return old.val_;    // already read this object

     const map<String_, reader_t>& readers = TheReaders();
     auto pr = readers.find(src.Type());
     REQUIRE(pr != readers.end(),
10           "No reader for '" + src.Type() + "'");
     Handle_<Data_> retval((pr->second)(src));
     src.Contents() = retval;
     return retval;
}
```

5.3   Rebuilding

Now that we have extracted our **Data_**, we must ask it to build the **Storable_** object in which we are really interested. Again, different concrete **Data_** types, with their disparate **Build** methods, are never gathered in one piece of code.

Building the **Data_** into a usable object follows a broadly similar path to extraction, but not identical. There are three main sources of difference:

- The **Build** method is declared in abstract classes which may have several concrete subclasses, so there are far fewer builders than readers. Backward compatibility constraints may further increase the number of readers but not of builders.
- When an object builds its subobjects, it already knows the subobject type and the signature of the build method, so only top-level objects ever need to search for a builder.
- Data about what is already built must be passed around explicitly, rather than queried using **Contents**.

Thus we define **Achive::Built_** and a type for builders:

─────────────────────── *Storable.h* ───────────────────────
```
namespace Archive
{
    struct Built_ : map<const Data_*, Handle_<Storable_> >
    {   };

5
    typedef Handle_<Storable_>(*builder_t)
          (const Handle_<Data_>&, Built_*);
}
```

There is a registry of builders, but it is simply a `vector<builder_t>` through which we will search linearly.

```
───────────────────── Storable.cpp ─────────────────────
Handle_<Storable_> Archive::Build
    (const Handle_<Data_>& data, Built_* built)
{
    if (built && built->count(data.get()))    // done
        return built->find(data.get())->second;

    const vector<builder_t>& bb = TheBuilders();
    for (auto pb = bb.begin(); pb != bb.end(); ++pb)
    {
        Handle_<Storable_> retval((*pb)(data, built));
        if (retval.get())
        {
            if (built)
                (*built)[data.get()] = retval;
            return retval;
        }
    }
    throw Exception_("No builder for archived data");
}
```

This relies on the registered builders returning an empty handle if called with an inappropriate object (which they will be, repeatedly). Also, it works only for "standardized" build methods which create a `Handle_` to a subclass of `Storable_`: `Data_` types which build something different will not be able to register their builders. But the whole point of the generic `Archive::Build` is to handle data about which all we know is that is corresponds to a storable handle; thus we can never register a nonstandard builder anyway.

5.3.1   *Some Syntactic Sugar*

The build process for a large object involves the building of many subobjects, probably across several layers of inclusion. The build record above is a key part of this process, but leads to bulky and illegible code. It is well worth fixing this up to support terse build methods.

We will overload the | operator (*not* ||) to correctly interpret code like `built | x` as "if x is already built, take the built result, otherwise build x." Data types with nonstandard built types (*i.e.*, not `Handle_s` of `Storable_s`) cannot use the build record, but we let them follow the same syntax. We accomplish this with partial specialization:

```
                          ── Storable.h ──────
   // generic no-op for non-handle types
   template<class T_, class BUILT_> void AddToRecord
       (Built_* record, const T_* src,
       const BUILT_& object) {}
 5 template<class BUILT_> void PopulateFromRecord
       (BUILT_* dst, const Handle_<Storable_>& src)
   { UNREACHABLE; }    // never called but must parse OK

   // partial specializations to handles
10 template<class DATA_, class BUILT_> void AddToRecord
       (Built_* record, const DATA_* src,
       const Handle_<BUILT_>& object)
   {
       if (record && object.get())
15     {
           record->insert(make_pair(src, HandleCast
               <Storable_>(object)));
       }
   }
20 template<class BUILT_> void PopulateFromRecord
       (Handle_<BUILT_>* dst, const Handle_<Storable_>& src)
   { *dst = HandleCast<BUILT_>(src); }

   // a sweet operator overload
25 template<class T_> typename T_::built_t operator|
       (Built_* built, const T_& data)
   {
       if (built && built->count(&data))
       {
30         // it's a handle, already built
           T_::built_t retval;
           PopulateFromRecord(&retval, (*built)[&data]);
           return retval;
       }
35     const T_::built_t& retval = data.Build(built);
       AddToRecord(built, &data, retval);
       return retval;
   }
```

We can further overload the operator to handle **Handle_** types, and optional data stored using **Maybe_**:

```
                          ── Storable.h ──────
   template<class T_> typename T_::built_t operator|
```

```
     (Built_* built, const Handle_<T_>& data)
   {
     return data.Empty() ? T_::built_t() : built | *data;
5  }
   template<class T_> typename T_::built_t operator|
     (Built_* built, const Maybe_<Handle_<T_> >& data)
   {
     return data.Known()
10           ? (built | data.Value())
             : T_::built_t();
   }
```

Thanks to Koenig lookup, we can define all these versions of **operator|** in **namespace Archive**. Finally, we add a function object for use in building several of one type (with **Apply**).

———————— *Storable.h* ————————

```
template<class T_> struct Build_
   : unary_function<Handle_<T_>, typename T_::built_t>
{
   Built_* built_;
5  Build_(Built_* built) : built_(built) {}
   typename T_::built_t operator()
       (const Handle_<T_>& data) const
   { return built_ | data; }
};
```

For instance, we might write

```
auto myTrades = Apply(Build_<TradeData_>(built), trades);
```

5.4   Code Generation

The code displayed in Sec. 5.2.2, particularly the archive reader, is verbose and error-prone to write. This is because it is a low-level representation of high-level concepts. We can represent the same information in a clear and compact form:

————————— *Interp1Linear.1.storable.if* —————————

```
ISA Interp1
NUMBER[] x
'    The abcissas (independent variables values)
'    of the interpolation
5  NUMBER[] f
'    The function (dependent variable) values
```

```
'   of the interpolation
CONDITION {data->x.size() == data->f.size()} \
   {x and f must have the same size}
```

The .1 in the mark-up file name is a version label, which helps "future-proof" us against the contingency of later needing to store different data for the same purpose.

This provides enough information to define **struct Data_Interp1Linear_v1_**.

- The **ISA** keyword states that this struct will inherit from **Data_Interp1_** (defined elsewhere), including that class's definition of **built_t**.
- The type used in **Save**, like the **struct** name, is formed from the file-name.
- The **CONDITION** keyword introduces a validation test, and gives the message to show if it fails.
- The **ArchiveExtract** function follows a mechanical process for each member.
- Each line is a comment (explaining the previous entry), a keyword directive, or the declaration of a member.
- Some member types, like **NUMBER[]**, are recognized and treated specially; an unrecognized type, like **Interp1**, would create a member of type **Handle_<Data_Interp1_>** to be extracted with a utility function **Archive::Interp1**.

Thus we would still write the Build function from Sec. 5.2.2, but we would replace the rest of the code with

```
———————— Interp1inear.cpp ——— —————————
namespace {
//:INSERT\H Interp1Linear.1.storable
//:INSERT\C Interp1Linear.1.storable
}
```

Inserting the "header code" (really, the class definition) within the source file insulates other code from these implementation details. The machine-generated code will define in C++ a concrete **Data_** object, and will also register a reader for it. It must be supported by a way to define the abstract superclasses in the public type hierarchy. These are defined in more interface definition files, which contain just a single line:

```
──────────────── Interp1.abstract.storable.if ────────────────
BUILDS Handle_<Interp1_>
```

If an abstract `Data_` type builds a nonstandard type (one not convertible to `Handle_<Storable_>`), we will mark the fact by writing `BUILDS!` rather than `BUILDS`. Between the file name and this one line, we now have enough information to define the abstract type `Data_Interp1_` and the utility function

```
──────────────── Interp1.cpp ────────────────
  Handle_<Data_Interp1_> Archive::Interp1
     (_ENV, const Src_* src)
  {
     assert(src != 0);
5    Handle_<Data_Interp1_> retval(HandleCast
           <Data_Interp1_>(Read(_env, *src)));
     REQUIRE(!retval.Empty(), "Could not read Interp1");
     return retval;
  }
```

This will be called by any `Data_` which contains an `Interp1` member.

A "nonlocal" error in the mark-up is one that cannot be detected from analyzing a single mark-up block, such as a concrete type whose `ISA` refers to a nonexistent base type. Such errors will lead to C++ compilation errors, or occasionally link errors. This makes sense since each mark-up block is its own translation unit.

At this point, persistence requires only two pieces of handwritten C++ code: the `Build` method which brings us back from the world of archives and data to that of working objects, and the implementation of `Save` in `Storable_` objects which sends us there in the first place. The latter is very similar to the machine-generated `Save` code for the corresponding `Data_` object. We also catch most spelling errors by defining names using the macro

```
──────────────── DA.h ────────────────
  #define BAREWORD(word) static const String_ word(#word);
```

Even so, the most obvious strategies each have drawbacks:

- Hand-write the `Storable_::Save` method – this is boring and error-prone.
- Cut-and-paste the `Data_::Save` method to provide its body – this removes the boredom without removing the danger.

- Have concrete `Storable_` objects hold a `Handle_` to a `Data_` object, and invoke its `Save` – this artificially constrains the authors of most objects.

To do better, we machine-generate a free function which will be called by the `Save` methods for both `Storable_` and `Data_` objects.

```
                          InterpLinear.cpp
    namespace Interp1Linear_v1
    {
        namespace Names
        {
  5         static const char* type_ = "Interp1Linear_v1";
            BAREWORD(x);
            BAREWORD(f);
        }

 10     void Save(Archive::Dst_& dst,
            const Vector_<>& x, const Vector_<>& f)
        {
            dst.SetType(Names::type );
            dst.Child(Names::x) = x;
 15         dst.Child(Names::f) = f;
        }
    }
    struct Data_Interp1Linear_v1_ : Data_Interp1_
    {
 20     Vector_<> x;
        Vector_<> f;
        built_t Build(Archive::Built_* built) const;
        void Save(Archive::Dst_& dst) const
        {
 25         Interp1Linear_v1:.Save(dst, x, f);
        }
    };
    Data_Interp1_::built_t Data_Interp1Linear_v1_::Build
        (Archive::Built_*) const
 30 {
        return Handle_<Interp1_>(Interp::NewLinear1(x, f));
    }
```

The code for `Build` here is unchanged from Sec. 5.2.2. Of course, we would make similar changes in the `ArchiveExtract` function, which is not displayed here. These defined names could then replace the quoted strings in the handwritten `Save` methods.

There is a complication in the free function helper for **Save** for composite objects. A **Data_** object contains members of type **Handle_<Data_>** (or a derived class), while a **Storable_** object probably contains **Handle_<Storable_>**. Thus the machine-generated implementations must be template functions, with a template type parameter for each child object:

```
————— PdeCoordinate.cpp —————
   namespace PdeCoordinateVector_v1
   {
      template<class T_> void Save
         (Archive::Dst_& dst,
 5          double y_low,
            double y_high,
            Date_ num_points,
            const T_& mapping)
      {
10       dst.SetType(Names::type_);
         dst.Child(Names::y_low) = y_low;
         dst.Child(Names::y_high) = y_high;
         dst.Child(Names::num_points) = num_points;
         dst.Child(Names::mapping) = mapping;
15    }
   }
```

Here **mapping** has a different type depending on whether we are storing our own **PdeCoordinateVector_** or its **Data** type; but our syntactic sugar ensures that **operator=** will always have the desired effect. We could simplify the code's appearance still further with the macro

```
#define WRITE_CHILD(x) dst.Child(Names::x) = x
```

but this is a false economy; the code will be machine-generated anyway and the macro simply hides the final C++ code from human eyes.

In practice, this implementation almost entirely eliminates the use of handwritten code to manipulate the archive; thus the **Names** offer a largely redundant safeguard and might be dropped.

Note that the machine-generated code is centered around the machine-generated **Data** type; we hand-write the **Save** and **Build** functions (though the former will likely be a straightforward call to a machine-generated function) which cross to and from the world of data. This preserves implementation freedom for the working classes.

5.5   A Display Interface

A particular archive of interest is one which converts (*"splats"*) a `Storable_`
to a two-dimensional tableau, which can then be displayed in a spreadsheet
or stored as a tab-separated file. For this we need concrete implementations
of `Archive::Dst_` and `Archive::Src_`; we will (almost) call them `Splat_`
and `Unsplat_` respectively.

5.5.1   *Storage*

A `Splat_` will, once fully populated, convert itself to a two-dimensional
table. The table's size will be that of what it stores (if it contains a concrete
datum), or the combined sizes of its children or elements.

Consider the virtual members of `Archive::Dst_` (in Sec. 5.1). The
overloaded versions of `operator=` will set internal data to contain the right-
hand side of the assignment. All these concrete types can be mapped onto a
two-dimensional `Matrix_<Cell_>`, so this is the type we will use for storage.
(A `Dictionary_` can be represented as a two-column array of keys and
values, or as a string.)

Alternatively, the archive might have to contain a `Storable_` composite
object. In this case, it will receive a type and some set of `Child` or `Element`
data, for each of which it will create a new `Splat_`.

Finally, the function `StoreRef` depends on information shared across
the entire archive. We can implement this as a member `shared_ptr`, created by the parent archive and passed to children, or as data outside the
class to which each `Splat_` object receives a reference. The latter approach
is clearer and more concise, but we must ensure that the reference is valid
as long as it is needed, so we can never let a `Splat_` escape the reference's
scope. Thus the class name is actually `XSplat_`, and it is locally names-
paced within a source file:

```
───────────────── Splat.cpp ─────────────────
class XSplat_ : public Archive::Dst_
{
    map<const Storable_*, String_>& refs_;
    String_ ownRef_;
    // String_ dataType_;
    Matrix_<Cell_> data_;
    String_ type_;    // if composite
    map<String_, shared_ptr<XSplat_> > children_;
```

A concrete value is stored into our own `data_` table, while composite values declare a type and then are stored as `children_`. Elements are just children with numeric names.

```cpp
————————————————— Splat.cpp ——————————————

   Dst_& Child(const String_& name)
   {
       shared_ptr<XSplat_>& ret = children_[name];
       assert(!ret.get());
5      ret.reset(new XSplat_(refs_));
       return *ret;
   }
   Dst_& Element(int i)
   {
10     return Child(String::FromInt(i));
   }
```

Assignment to a concrete type creates the array of values for storage:

```cpp
————————————————— Splat.cpp ——————————————

   void operator=(const Matrix_<Cell_>& val)
   {
       // dataType_ = "CELL[][]";
       data_ = val;
5  }
   void operator=(double val)
   {
       // dataType_ = "NUMBER";
       data_.Resize(1, 1);
10     data_(0, 0) = val;
   }
```

and so on through the other concrete types. The `dataType_` is optional; it can be used to make distinctions like the one shown here, or we can simply treat all data as one type. We prefer the latter approach, and display the option only for completeness.

A reference is a unique identifier for an already-written object, but our scheme requires us to form it for an object about to be written.

```cpp
————————————————— Splat.cpp ——————————————

   bool StoreRef(const Storable_& object)
   {
       bool retval = refs_.count(&object) != 0;
       if (!retval)
5          refs_[&object] = UniqueName(object);
       ownRef_ = refs_[&object];
       return retval;
```

```
      }
      bool StoreRef(const Handle_<Storable_>& object)
10    {
          return StoreRef(*object);
      }
};
```

Thus failure to find a stored reference causes a reference to be stored for the future. This dovetails with our requirement that objects must write their contents iff **StoreRef** returns **false**.

To create the final output, we must decide on a format for the table. We use the first column to distinguish between fundamental and composite objects: it will contain the string "DATA" (or the **dataType_**, if used) for the former, and a special character like ~ followed by the **type_** for the latter. The next cell in the first column will be used for composite types to contain the reference generated by **UniqueName**. A couple of supporting functions will prove convenient:

```
––––––––––––––––––– Splat.cpp ––––––––––––––
Matrix_<Cell > M1x1(const String_& src)
{
    Matrix_<Cell_> retval(1, 1);
    retval(0, 0) = src;
5   return retval;
}
Matrix_<Cell_> SideBySide
    (const Matrix_<Cell_>& m1,
    const Matrix_<Cell_>& m2)
10 {
    return Matrix::Merge(String_("1,2"), Join(&m1, &m2));
}
```

Here **M1x1** creates a 1×1 matrix whose single cell contains the input **String_**. **Join** and **Merge** are discussed in Ch. 4. Then we support the writing process with a new member function of **XSplat_**:

```
––––––––––––––––––– Splat.cpp ––––––––––––––
Matrix_<Cell_> XSplat_::Write() const
{
    assert(data_.Empty() != type_.empty());
    if (type_.empty())   // fundamental
5   {
        String_ tag("DATA");   // or dataType_
        return SideBySide(M1x1(tag), data_);
    }
```

```
    // composite
10  Matrix_<Cell_> labels;
    Matrix::Append(&labels, M1x1('~' + type_));
    Matrix::Append(&labels, M1x1('^' + ownRef_));
    Matrix_<Cell_> childVals;
    for (auto pc = children_.begin();
15         pc != children_.end(); ++pc)
    {
        Matrix::Append(&childVals, SideBySide
            (M1x1(pc->first), pc->second->Write()));
    }
20  return SideBySide(labels, childVals);
}
```

To address the scope-of-reference problem mentioned above, this class is defined in a local namespace and accessed only through the function

Splat.cpp

```
Matrix_<Cell_> Archive::Splat(const Handle_<Storable_>& ob)
{
    map<const Storable_*, String_> refs;
    XSplat_ dst(refs);
5   Archive::Write(dst, ob);
    return dst.Write();
}
```

5.5.2 *Extraction*

The `Unsplat_` reader must decipher the table thus created, or more generally, a `SubMatrix_` of the table. By looking at the first cell, the reader can tell whether it contains a fundamental or a composite type. The different fundamental types are created by supporting functions which return a default-constructed `Maybe_` in case of failure. For example:

Splat.cpp

```
typedef Matrix_<Cell_>::SubMatrix_ section_t;

Maybe_<double> XAsNumber(const section_t& src)
{
5   if (src.Rows() == 1 && Cell::IsNumber(src(0, 0)))
    {
        Matrix_<Cell_>::ConstColumn_ col = src.Column(0);
        if (find_if(col.begin() + 1, col.end(),
                not1(Cell::TypeCheck_().Empty())) == col.end())
10      {
            return Cell::AsNumber(src(0, 0));
```

```
          }
      }
      return Maybe_<double>();
15 }
```

This tests a section of the displayed object to see if it might be a single number: if so, it will have one row, the remainder of which will be blank. This supports the `AsNumber` member function in our reader:

```
————————————— Splat.cpp ——————————————
class XUnsplat_ : public Archive::Src_
{
    section_t vals_;
    map<String_, Handle_<Archive::Data_> >& refs_;

5
public:
    XUnsplat_(const section_t& vals, map<String_,
          Handle_<Archive::Data_> >& refs)
    : vals_(vals), refs_(refs) {}

10
    bool IsComposite() const
    {
        assert(!vals_.Empty());
        assert(Cell::IsString(vals_(0, 0)));
15      return Cell::AsString(vals_(0, 0))[0] == '~';
    }

    Maybe_<double> AsNumber() const
    {
20      return IsComposite()
          ? Maybe_<double>()
          : XAsNumber(vals_.SubMatrix(0, vals_.Rows(), 1));
    }
```

If we are using the `dataType_` from the previous section, then the implementation must check it before proceeding. Other `As` functions will be implemented along the same lines.

To get a list of `Children`, we walk down the next-to-leftmost column (after checking that the leftmost column contains the expected type and reference information). Each nonblank entry in this column is the name of a child, whose data live in the corresponding submatrix.

```
————————————— Splat.cpp ——————————————
    map<String_, Handle_<Src_> > Children() const
    {
        map<String_, Handle_<Src_> > ret;
```

```
     if (IsComposite())
     {
         auto begin = vals_.Column(1).begin();
         if (begin->empty())
             return ret;    // no child data
         int previous = 0;
         while (previous < vals_.Rows())
         {
             assert(NonBlank(vals_(previous, 1)));
             auto stop = find_if(begin + previous,
                 vals_.Column(1).end(), NonBlank);
             const int next = stop - begin;
             const String_ name = Cell::AsString
                     (vals_(previous, 1));
             REQUIRE0(!ret.count(name),
                 "Duplicate child '" + name + "'");
             ret[name].reset(new XUnsplat_
                     (vals_.SubMatrix
                             (previous, next, 2),
                         refs_));
             previous = next;
         }
     }
     return ret;
}
```

Elements is implemented in terms of **Children**; the version here enforces our expectation that elements should be continuously numbered from zero.

```
————————————————— Splat.cpp ——————————
     vector<Handle_<Src_> > Elements() const
     {
         map<String_, Handle_<Src_> > cm = Children();
         vector<Handle_<Src_> > ret(cm.size());
         for (auto pc = cm.begin(); pc != cm.end(); ++pc)
         {
             if (String::IsNumber(pc->first))
             {
                 const int ic = String::ToInt(pc->first);
                 REQUIRE0(ic >= 0 && ic < ret.size(),
                     "Child index '" + pc->first +
                     "' is out of range");
                 ret[ic] = pc->second;
             }
         }
         while (!ret.empty() && ret.back().Empty())
```

```
        ret.pop_back();
        Require(0, FindIf(ret, IsEmpty) == ret.end(),
          "Missing children");
20      return ret;
    }
```

The contents is a reference to an element in the map of references:

```
────────────────────────── Splat.cpp ──────────────────
    Handle_<Archive::Data_>& Contents() const
    {
        assert(IsComposite());
        auto tag = Cell::AsString(vals_(1,0)).substr(1);
5       return refs_[tag];
    }
};
```

This relies on the **XSplat_** writer's use of a single-character prefix to identify the reference tag.

We can choose how to implement **Unexpected**, depending on the desired tradeoff between forgiving flexibility and bulletproof reliability.

Because **XSplat_** again depends on a reference to a non-member (**refs_** above), we create it in a local namespace and access it only through a utility function:

```
────────────────────────── Splat.cpp ──────────────────
Handle_<Storable_> Archive::Unsplat
    (_ENV, const Matrix_<Cell_>& src)
{
    map<String_, Handle_<Archive::Data_> > refs;
5   XUnsplat_ reader(src.SubMatrix(0, 0), refs);
    Handle_<Data_> parsed = Read(_env, reader);
    Built_ record;
    return Build(parsed, &record);
}
```

5.5.3 *Refinements*

This describes a minimal object display engine; several refinements will likely be desirable.

- Allow specification of an *element path* so that a sub-object, rather than the entire object, can be viewed.
- Add a tab-separated file reader/writer to convert the **Matrix_** data to files; or generalize the data type used to support both matrices and

files.

- Allow truncation of the output at a user-input tree depth, so that objects can be inspected gradually from the top.
- Interact with the repository, replacing child data (beyond the specified depth) with a repository handle (which can be splatted in turn, if desired).

Since the table output from `Splat` can be written to a file, it can serve as the sole archive method for many purposes. However, there is a widespread feeling that XML files, or some similar format with third-party backing, are more "official" and more suited for books and records.

5.6   Auditing

During a long computation, we will form and discard complicated objects; for example, a risk run usually involves the creation of temporary bumped models. Inspection of these objects can help greatly in debugging or validating a process; but they are no longer with us.

We would like to have the option to preserve these objects. Clearly they cannot be carried out along the call stack.[2] But we can store them in our environment.

5.6.1   *Bag*

The first requirement is a place to store objects. We will require that they be `Storable_`, since a non-storable object is of little utility outside its local context; it cannot be put in the repository, or displayed, or written to a file. We need a key facility to attach additional information and to search for a given piece; but the keys may not be unique. Thus we arrive at

```
───────────────── Bag.h ─────────────────
class Bag_ : public Storable_
{
public:
    multimap<String_, Handle_<Storable_> > vals_;
5   scoped_ptr<RepositoryErase_> rPolicy_;

    Bag_(const String_& name) : Storable_("Bag", name) {}
};
```

---

[2]It is an interesting exercise to enumerate the distinct ways in which this is infeasible.

When reconstituted from an archive, a `Bag_` with a repository policy will write its contents to the in-process repository. This lets us use a `Bag_` as a container of environment if desired. We might supply no policy, meaning that the repository is not to be accessed. `Maybe_<RepositoryErase_>` is problematic because we prefer that `RepositoryErase_` should lack a default constructor, so we use an empty `scoped_ptr` instead. Naturally, `RepositoryErase_` is implemented as a machine-generated enumerated class; see Sec. 3.8.

The public interface of `Bag_` must allow us to view, insert and erase objects; thus we simply make `vals_` public data. We could make `rPolicy_ const` and require it as an input to the constructor; but this would make the class more confusing, because of the need to specify the fate of the input pointer (is its memory captured by the constructor)? It is simpler to populate the policy after making the `Bag_`.

5.6.2 *Filling Up*

We need an auditor, an object which can be passed within the environment. Functions which create temporary `Storable_` objects – *i.e.*, not their return values – can and generally should put them in handles and show them to the auditor. We may go out of our way to make temporary objects `Storable_` for this purpose.

The auditor may also be used to communicate between successive iterations of a complex algorithm (*e.g.*, between base and bumped valuations). Thus we provide a query interface to check for the presence of some named object.

```
───────────────────────── Audit.h ─────────────────────────
class Auditor_ : public Environment_::Entry_
{
public:
    virtual void Notice
        (const String_& key,
          const Handle_<Storable_>& value)
    const = 0;

    virtual Vector_<Handle_<Storable_ > > Find
        (const String_& key)
    const = 0;
};
```

We will support these with `namespace Environment` utilities

```
                        ─── Audit.h ───
template<class T_> void Audit
   (_ENV, const String_& key, const Handle_<T_>& value)
{
   AuditBase(_env, key, HandleCast<Storable_>(value));
5 }
```

```
                      ─── Audit.cpp ───
void Environment::AuditBase(_ENV,
   const String_& key, const Handle_<Storable_>& value)
{
   ShowToAuditor_ f(key, HandleCast<Storable_>(value));
5  Environment::Iterate(_env, f);
}
```

where **ShowToAuditor_** is a function object which checks (using **dynamic_cast**) whether the input **Environment::Entry_** is an **Auditor_**, and if so calls **Notice**. The names must be chosen to show the direction of information flow, so we avoid words like "remember" – which means either to store in, or to recall from, memory – and "show," since it is not clear whether an object is being shown to or by the auditor.[3]

User code, once it has a handle **localThing** to whatever object it creates, simply calls

```
Environment::Audit(_env, "name", localThing);
```

The common use case of **Find** is to fetch forth a single **Notice**d object, and place it in an empty handle:

```
                        ─── Audit.h ───
template<class T_> struct Recall_
{
   const String_ key_;
   Handle_<T_>* value_;
5  const Environment_* env_;
   Recall_(_ENV, const String_& key, Handle_<T_>* value)
      : env_(_env), key_(key), value_(value) {}
   void operator()(const Entry_& env) const
   {
10     if (DYN_PTR(audit, const Auditor_, &env))
       {
          auto fh = audit->Find(key_);
          for (auto qh = fh.begin(); qh != fh.end(); ++qh)
```

─────────────────────────

[3]The similarity of this name to our **NOTICE** macro is largely deliberate, and there is no chance to confusing the two.

```
15                  {
                        Handle_<T_> temp = HandleCast<T_>(*qh);
                        if (!temp.Empty() && temp != *value_)
                        {
                            Require(env_, value_->Empty(),
                                "Conflicting recollections");
20                          *value_ = temp;
                        }
                    }
                }
            }
25  };

    template<class T_> void Recall
        (_ENV, const String_& key, Handle_<T_>* value)
    {
30      assert(value && value->Empty());
        NOTE(key);
        Iterate(_env, Recall_<T_>(_env, key, value));
    }
```

This somewhat awkward code supports a reasonably compact usage idiom:

```
    Handle_<MyType_> local;
    Environment::Recall(_env, "theKey", &local);
    if (local.Empty())
    {
5       // make it ourselves
        // ...
        Environment::Audit(_env, "theKey", local);
    }
```

Sec. 14.7.1 shows a realistic example. This provides a context with which a computation can detect that it is a bumped case (a perturbation), and have some chosen data available from the base computation. We can use this as a generic stabilization method for perturbations; see Sec. 14.4.

5.6.3 *Audit Types*

Auditors are not obliged to show all their contents to any caller of Find; also note that we provide no facility to find all the keys. The expected implementation of Auditor_ holds a Bag into which it might place objects:

```
                            Audit.cpp
struct AuditorImp_ : Auditor_
{
    shared_ptr<Bag_> mine_;
    enum
    {
        PASSIVE,
        READING,
        READING_EXCLUSIVE,   // avoid vast memory use
        SHOWING,
    } mode_;
    // ...
};
```

The default constructor will set the `mode_` to `READING`, meaning that the auditor should hang onto every object sent to `Notice`.

```
                            Audit.cpp
void AuditorImp_::Notice
    (const String_& key,
    const Handle_<Storable_>& value)
const
{
    switch (mode_)
    {
    case READING_EXCLUSIVE:
        mine_->vals_.erase(key);    // and fall through
    case READING:
        mine_->vals_.insert(make_pair(key, value));
        break;
    }
}
```

Other modes ignore those objects but will **Find** their stored objects:

```
                            Audit.cpp
Vector_<Handle_<Storable_> > AuditorImp_::Find
    (const String_& key)
const
{
    static Get2nd_<String_, Handle_<Storable_> > getV;
    Vector_<Handle_<Storable_> > retval;
    if (mode_ == SHOWING)
    {
        auto range = mine_->vals_.equal_range(key);
        transform(range.first, range.second,
                back_inserter(retval), getV);
```

```
    }
    return retval;
}
```

An `AuditorImp_` belongs to its creator, who alone can access its full
data (except by sneaky casting, easily found by a global code search) or
change its mode.

To audit a given function, we add an empty `AuditorImp_` to the environment before the call, and later pick up its `Bag_` of data. For a toy
example, consider the helper function `Interp1_Get` from Sec. 3.2; we can
write

```
Handle_<Bag_> Interp1_Get_Audit(_ENV,
    const Interp1_& f, const Vector_<>& x, Vector_<>* y)
{
    AuditorImp_ audit;
    Environment::XEphemeral_ envA(_env, audit);
    Interp1_Get(&envA, f, x, y);
    return audit.mine_;
}
```

A more general auditing mechanism is best implemented as part of a larger
project, creation of a scripting language, which is outside the scope of this
volume.

5.7   More on Repositories

Another use of `Bag_` is to hold the entire state of an interactive session.
Since the only form of state which persists between function calls is that of
our repository,[4] which holds only `Storable_` objects, a `Bag_` can be loaded
up with the entire repository and saved. A later session reads in the `Bag_`,
which causes its contents to be added the that session's repository; see
Sec. 5.6.1.

5.7.1   *Unique Objects*

We choose to make the distinction between past and future part of the
environment, rather than provide it as a function input, since all valuation
and risk functions need this information. This distinction is discussed in

---

[4]One of our tasks is to make this so.

detail in Sec. 10.1; here we consider the storage of the *accounting date* which separates future payments from those already paid.

Upon initialization of a session, the repository will be empty, and we will initialize a system default date (probably the current local day at load time). We allow this to be overridden by storing a repository object (to be defined locally to a C++ source file):

```
───────────── GlobalDates.cpp ─────────────
    struct AccountingDateOverride_ : Storable_
    {
        Date_ date_;
        AccountingDateOverride_(Date_ date)
5           :
        Storable_("AccountingDateOverride", String_()),
        date_(date)
        {   }

10      void Save(Archive::Dst_& dst) const
        {
            Save_AccountingDateOverride_v1(dst, date_);
        }
    };
```

This object has no name, because we never want to support multiple instances simultaneously; its constructor reflects this.

Now we set the accounting date by storing a new `AccountingDateOverride_`, which will overwrite any others in the repository.

```
───────────── GlobalDates.cpp ─────────────
    void Environment::Change::SetAccountingDate(_ENV, Date_ d)
    {
        Handle_<Storable_> s(new AccountingDateOverride_(d));
        (void) Repository::Add(s, RepositoryErase_::TYPE);
5   }
```

The query function checks for the existence of such an object (see Sec. 3.7.4) and, if none is found, uses the system default stored at start-up.

We allow temporary manipulation of such environment variables using the RAII idiom. We put this class inside `namespace Environment`:

```
───────────── GlobalDates.h ─────────────
    struct TemporaryAccountingDate_ : noncopyable
    {
        Date_ save_;
        TemporaryAccountingDate_(_ENV, Date_ override);
```

```
5    ~TemporaryAccountingDate_();
  };
```

The constructor and destructor are short and simple, but there is no gain from inlining them.

```
─────────────── GlobalDates.cpp ───────────────
TemporaryAccountingDate_::TemporaryAccountingDate_
    (_ENV, Date_ override)
  : save_(Environment::AccountingDate())
  {
5    Environment::Change::SetAccountingDate(_env, override);
  }
TemporaryAccountingDate_::~TemporaryAccountingDate_()
  {
    Environment::Change::SetAccountingDate(0, save_);
10 }
```

5.7.2 *Naming*

To store objects in our in-process repository, we must decide how to name them; we need to specify how "long names", which will be used as repository keys, will be generated for a given object. Each `Storable_` has a `type_` and `name_`, which should both be included in the long name; but we must also provide a way to distinguish successive instances with the same type and name (e.g., a yield curve which is continually updated with new market data). Of course, the long name must be unique within an interactive session; thus we must attach extra text within the long name. Several approaches are possible:

- A version number or "ticker", incremented each time a new object is created;
- Variants where a separate ticker is used for each type, or each name, or each combination thereof;
- A string representation of the object's address in memory;
- A GUID (128 bits, 32 hex chars, 20-26 printable chars);
- The chip-independent part of a GUID ($\frac{5}{8}$ as big).

The first two approaches seem more friendly to the user, who sees each object with a numeric version rather than an unreadable multi-character "dongle." However, this is actually a disadvantage if it invites the user to type long names by hand, thus breaking the dependency chain of a calculation. We prefer the third approach, which can be implemented with about seven characters of extra information.

Long names should also be immediately identifiable as such. For this purpose it is worthwhile to reserve a character like ~ or &, which will not naturally be used in object naming, and use it as a separator; thus long names will look like ~YC~USD1~g8h6w04.

5.7.3 *Matching*

Fetching forth an object's long name from incomplete information is not always a mistake. For instance, a pricing spreadsheet may be intended to exist independently of any single yield curve, and to operate using whatever curve (with the appropriate currency) is present in the environment. Thus we must create a function to search the repository and return long names.[5]

This is best done with *pattern matching*. Rather than just supply the beginning of the name, the user gives a pattern to match, and we return all long names matching it. So matching ~YC~ is a search for all yield curves,[6] while matching ~YC~USD1~ is a search for all yield curves with that particular name. While Perl (or, even better, Perl 6) regular expressions are the gold standard, we can make do with much simpler patterns. Supporting ^, which matches only the front of a string, is a useful optimization, permitting us to restrict the set of repository keys which must be tested.

Users will control the names given, and can adapt their searches to their own naming conventions. A couple of refinements are useful:

- Simultaneous match of multiple patterns, both "and" and "or".
- An optional flag to enforce uniqueness, returning an error unless exactly one match is found.
- Permitting ^ to consume the leading ~ or other special character.

This function lets users examine the repository, to find an object or to fetch all objects (*e.g.*, for creating a bag of environment).

---

[5] This assumes a desktop-based computation model; cloud computation poses different challenges.

[6] And also, unless we change our naming system, for other objects which happen to be named "YC".

Chapter 6

Testing Framework

Like any large software project, our library will evolve through time and will need constant upkeep to remain reliable. In attempting to do more tasks with less code, we are inevitably forcing code to be used in many different contexts, so a bug will often only be manifest in some uncommon or unexpected use case. Also, testing of the interaction between high-level components requires a great deal of low level work, even if the low-level components are known to be reliable.

6.1 Component Tests

Inside the code, we can add tests of individual functions or classes. These tests are themselves functions, which can in theory be the `main` functions of many individual executable tests.[1] This decreases the overhead of selecting a specific test, at the expense of increased maintenance and link time. We prefer to run all tests automatically, rather than rely on the developer's judgement of which are needed. This leads us to the opposite approach: a single large test executable, which calls separately written test functions in sequence. We will provide the test executable with a singleton registry of functions to call, and use our usual registration idiom:

Test.h

```
namespace Test
{
    typedef void(*func_t)();
    void Register(func_t func);
    void Fail(const char* msg);
    void Fail(const String_& msg);
}
```

---

[1]This approach seems to be implicit in Lakos's *Large-Scale C++ Software Design*.

91

```
      struct ComponentTest_
10    {
          ComponentTest_(Test::func_t fn) { Test::Register(fn); }
      };
      #define TESTFUNC(nm) \
      void nm();           \
15    static const ComponentTest_ test_register__##nm(nm); \
      void nm()
```

This definition of TESTFUNC lets us use it as a function signature directly in the code:

NCDF.test.cpp

```
      TESTFUNC(TestNCDF)
      {
          using SpecialFunctions::N;
          TEST(N(0.0) == 0.5);
5         // ...
```

Here we have introduced a new macro, which we implement as

Test.h

```
      #define TEST(cond) if (cond); else Test::Fail(#cond)
```

The function `Test::Fail` implements the consequences of failure; most likely it will simply display a message to `cout` and increment an error count, but more sophisticated behaviors are possible.

Since a large part of our work is numerical, we will implement some other macros to streamline numerical tests.

Test.h

```
      #define TEST_SIMEQ(x, y, eps)    \
      {const double myX = x, myY = y;        \
      if (fabs(myX - myY)       \
              <= eps * Max(1.0, Max(fabs(myX), fabs(myY)))); \
5     else Test::Fail(#x "~" #y "(within " #eps ")")

      #define TEST_EQ(x, y) TEST_SIMEQ(x, y, DA::EPSILON)
```

The input tolerance `eps` is thus interpreted as a relative tolerance for large numbers, and as an absolute tolerance for small numbers. This is the test we most often desire; we can easily implement other macros like `TEST_ABSOLUTE` and `TEST_RELATIVE` as needed.

Another macro tests for the emission of an error message:

Test.h

```
#define TEST_ERROR(x, m) \
try {x; Test::Fail(#x " did not generate the error");} \
catch (Exception_& e) {TEST(e.Display() == m);}
```

6.1.1 *Physical Structure*

We are reluctant to put tests directly into a source file, since they clutter the implementation. But the test code often needs access to the full implementation of a class, not just the public interface; and keeping the implementation insulated in the source file is itself an important design aim.

Our solution is to include *the source itself* within a testing file. For example, we might create NCDF.test.cpp which would begin

---
NCDF.test.cpp

```
#include "NCDF.cpp"
#include "ComponentTest.h"

TESTFUNC(TestNCDF)
```
---

and go on to implement other testing functions.

We then create two separate builds, incorporating either NCDF.test.cpp or NCDF.cpp; these build the library with or without component tests. The latter is more suitable for release into production.

The test executable implements Test::Register – thus, during the loading of the library (with component tests), each test declared using TESTFUNC will be registered. If we wish to load the library with component tests into a different executable, we must provide a stub version of Test::Register.

6.1.2 *Reuse*

Creating a high-level object for testing requires many prerequisite objects: for instance, to test a risk computation, we need a model, which needs a yield curve, which needs instruments. This threatens to make testing code cumbersome to the point of uselessness.

We must add additional functionality, in the test files, to create lower-level objects which can be reused when testing high-level objects. It would be nice to reuse the auditing tools from Sec. 5.6 to store these objects, rather than building them from scratch for each test; unfortunately, that would require tracking the dependencies between component tests. Thus we stick with the low-tech solution of simply defining helper functions and declaring them in test header files for later use. A typical example is

```
————————————— YcBuild.test.h —————————————
Handle_<YieldCurve_> Test::YC_GBP30Y(long quote_date)
```

6.2 Regression Tests

Tests which use only the public interface of the library need not be written in C++. We can take advantage of interactive environments, notably Excel, to more quickly put together such a test. These are more useful for regression testing, to flag unexpected changes, than for component correctness testing.

Sec. 5.7 describes how we can "bag up" the whole environment of an interactive session; thus a stored regression test can consist of a bag of objects, plus a continuation of the interactive session. For example, we might store a file containing the session state, and an Excel spreadsheet to be run after the file is loaded. The spreadsheet's cells, or some subset thereof (*e.g.*, those marked with a cell comment), could be compared across releases.

6.2.1 *Repository Instrumentation*

It is not necessary to bag the whole environment; if we mediate repository access, we can add code to record which objects were fetched during a computation. This process is known as *instrumentation*, in this case within `Environment::Fetch`. We rename the simple version `ObjectAccess_::Fetch` from Sec. 3.7.4 to `XFetch`, to highlight that it should not be called directly, and write

```
————————————— EnvironmentRepository.h —————————————
      template<class T_> Handle_<T_> Fetch
         (_ENV, const String_& tag, bool opt = false)
      {
         NOTICE("Handle tag", tag);
5        Handle_<Storable_> base = FetchBase(_env, tag, opt);
         Handle_<T_> retval = HandleCast<T_>(base);
         Require(_env, !retval.Empty() || base.Empty(),
              "Stored object is of wrong type");
         return retval;
10    }
```

```
─────────────── EnvironmentRepository.cpp ───────────────
   Handle_<Storable_> Environment::FetchBase
      (_ENV, const String_& tag, bool opt)
   {
      auto access = Extract1<ObjectAccess_>(_env);
5     Require(_env, opt || access, "No repository access");
      if (!access)
         return Handle_<Storable_>();
      else if (RepositoryListener_* listen = TheListener())
         return (*listen)(_env, *access, tag, opt);
10    else
         return access->XFetch<Storable_>(tag, opt);
   }
```

This requires a new class **RepositoryListener_**, stored as a singleton and accessed as **TheListener**.

One such listener simply records the handles returned:

```
─────────────── EnvironmentRepository.cpp ───────────────
   struct RepositoryRecording_ : RepositoryListener_
   {
      map<String_, Handle_<Storable_> > seen_;
      Handle_<Storable_> operator()
5        (_ENV, const ObjectAccess_& access,
          const String_& tag, bool optional)
      {
         Handle_<Storable_> retval
            = access.Fetch<Storable_>(tag, optional);
10       seen_[tag] = retval;
         return retval;
      }
   };
```

To "record" part of a session now means the following:

(1) Create an instance of **RepositoryRecording_** and set **TheListener** to it.
(2) Run the desired part of the session.
(3) Reset **TheListener** to 0.

After this process, the recording will have been populated with all the repository queries made.

In practice, repository instrumentation works best in combination with a scripting language, making it simple to turn a single execution of a script into a self-contained regression test.

6.3   No Silver Bullet

The ongoing challenge, of course, is to keep both component and regression tests up-to-date and comprehensive. Alas, this is still a matter of discipline, with no technological silver bullets on offer.

Chapter 7

Further Maths

Besides the vector and matrix computations discussed in Ch. 4, we require a good deal of other mathematical functionality, some of it quite specialized to the field of derivatives.

7.1 Interpolation

Our approach to construction of (one-dimensional) interpolants is unsurprising: we create an abstract base class `Interp1_` and derive concrete classes from it. Our only innovation is in cubic splines, where we support the ability to set the first, second or third derivative at each boundary. In namespace `Interp`, we write

```
                    ———————— InterpCubic.h ————————
struct Boundary_
{
    int order_;
    double value_;
    Boundary_(int o, double v) : order_(o), value_(v) {}
};
```

Thus `Boundary_(3, 0.0)` instructs the decomposition to set the third derivative to zero (I like to refer to this as the "super-natural" spline). We set the boundaries independently when splining:

```
                    ———————— InterpCubic.h ————————
Interp1_* NewCubic
    (const Vector_<>& x, const Vector_<>& f,
    const Boundary_& lhs, const Boundary_& rhs);
```

The factory function calls a constructor of a *local class*, *i.e.*, one defined only within the source file. The code for this is based on the `spline` routine of *Numerical Recipes*, except at the boundaries:

```
—————————————— InterpCubic.cpp ——————————
Cubic1_::Cubic1_(const String_& name, const Vector_<>& x,
    const Vector_<>& f, const Interp::Boundary_& lhs,
    const Interp::Boundary_& rhs)
: Interp1_(name), x_(x), f_(f), fpp_(f.size())
{
    assert(x.size() > 2 && IsMonotonic(x));
    assert(x.size() == f.size());
    const int n = x.size();
    Vector_<> u(n - 1);
    switch (lhs.order_)    // set left boundary
    {
    default:
        UNREACHABLE;
    case 1:
        {
            const double dx = x[1] - x[0];
            fpp_[0] = -0.5;
            u[0] = ((f[1] - f[0]) / dx - lhs.value_)
                * (3.0 / dx);
        }
        break;
    case 2:
        u[0] = lhs.value_;
        fpp_[0] = 0.0;
        break;
    case 3:
        u[0] = -(x[1] - x[0]) * lhs.value_;
        fpp_[0] = 1.0;
        break;
    }
    for (int i = 1; i < n - 1; ++i)    // decomposition
    {
        const double dx = x[i] - x[i - 1];
        const double d2 = x[i + 1] - x[i - 1];
        const double sig = dx /d2;
        const double p = sig * fpp_[i - 1] + 2.0;
        fpp_[i] = (sig - 1.0) / p;
        const double temp = (f[i + 1] - f[i]) / d2
            - (f[i] - f[i - 1]) / dx;
        u[i] = (6.0 * temp - dx * u[i - 1]) / (p * d2);
    }
    switch (rhs.order_)    // set right boundary
    {
    default:
```

```
45        UNREACHABLE;
       case 1:
          {
              const double dx = x[n - 1] - x[n - 2];
              const double un = (3.0 / dx) * (rhs.value_ -
50                   (f[n - 1] - f[n - 2]) / dx);
              fpp_[n - 1] = (un - 0.5 * u[n - 2])
                     / (1.0 + 0.5 * fpp_[n - 2]);
          }
          break;
55     case 2:
          fpp_[n - 1] = rhs.value_;
          break;
       case 3:
          {
60            const double dx = x[n - 1] - x[n - 2];
              fpp_[n - 1] = (u[n - 2] + dx * rhs.value_)
                     / (1.0 - fpp_[n - 2]);
          }
          break;
65     }
       for (int k = n - 2; k >= 0; --k)   // backsubstitution
       {
           fpp_[k] = fpp_[k] * fpp_[k + 1] + u[k];
       }
70  }
```

7.2   Special Functions

7.2.1   *The Normal Distribution*

The most important special functions in finance are the normal cumulative distribution $N(x)$ and its inverse.[1] The most widespread public-domain implementation is that of Abramowitz and Stegun from their *Handbook of Mathematical Functions*; but it is not terribly accurate, and several shops have built improved in-house approximations.

Before committing to this approach, we must distinguish between cases where high accuracy is needed, and those where speed is more important. It is quite likely that we could run an entire real-world operation using only a low-accuracy approximation – say, to four decimal places – of N; this

---

[1]Criticism of models based on normal distributions is widespread and justified, but these distributions also supply the foundation for more realistic models.

would be tantamount to using a slightly different distribution in pricing. The main difficulty would arise from lack of closure under convolution.[2]

Given an approximate value for $N^{-1}(x)$, such as the output of a low-precision analytical approximation, we can radically improve it by a single "polishing" using Newton's method. Suppose $y \sim N^{-1}(x)$, and let $y' \equiv y + (x - N(y))/\phi(y)$ where ϕ is the normal density. A single iteration usually suffices to achieve double-precision accuracy. Thus at the expense of one evaluation of N and one of ϕ, we can enforce consistency between our two approximations.

This leads us to the interface, in namespace SpecialFunctions,

```
————————————— NCDF.h —————————————
double N(double z, bool precise = true);
double NInverse
    (double x, bool precise = true, bool polish = true);
```

The defaults are chosen for reliability rather than speed. This is a good general practice: in known hotspots, or after profiling, we can switch over to the less-precise fast method, but we should never set it as the default.

Inside these functions, we will have a top-level switch based on the **precise** flag; the computational cost of this is insignificant. One candidate for an approximate N or NInverse is cubic spline on a precomputed set of points; it is possible to get absolute errors in N under 10^{-6} everywhere with as few as 16 spline points.[3]

```
—————————————— NCDF.cpp ——————————————
    static const int NCDFCS_N = 16;
    static const double NCDFCS_X[16] =
        {-3.734582185, -3.347382781, -3.030883722,
        -2.75090681, -2.492289824, -2.243141537,
5       -1.992179668, -1.494029881, -1.290815576,
        -1.120050999, -0.954303629, -0.792072249,
        -0.629093487, -0.460389924, -0.276889742, 0.0};
    static const double NCDFCS_F[16] =
        {9.47235E-05, 0.000408582, 0.001219907, 0.002972237,
10      0.00634685, 0.012444548, 0.023176395, 0.067583453,
        0.098383227, 0.131345731, 0.16996458, 0.214158839,
        0.264643073, 0.322617682, 0.39093184, 0.5};

    Interp1_* MakeNcdfSpline()
15  {
        static const Vector_<> x
```

---

[2]There is no pressing reason to try this experiment.

[3]In fact, this is one of the few sound applications of the cubic spline.

```
              (&NCDFCS_X[0], &NCDFCS_X[NCDFCS_N - 1]);
      static const Vector_<> f
              (&NCDFCS_F[0], &NCDFCS_F[NCDFCS_N - 1]);
20    const Interp::Boundary_ lhs(1, 0.000373538);
      const Interp::Boundary_ rhs(1, 0.39898679); // at x=0
      return Interp::NewCubic(x, f, lhs, rhs);
}
```

We will call this once to create a `static` spline interpolant:

```
——————————————————— NCDF.cpp ———————————————————
double NcdfBySpline(double z)
{
    static const scoped_ptr<Interp1_> SPLINE
            (MakeNcdfSpline());
5    if (z > 0.0)
        return 1.0 - NcdfBySpline(-z);
    return z < NCDFCS_F[0]
            ? NCDFCS_X[0] * exp(4.17528563* (z - NCDFCS_F[0]))
            : SPLINE->Get(z);
10 }
```

We choose to compute $N(x)$ at negative x in order to obtain accurate values when N is very small – even when $1 - N$ is indistinguishable from 1.

7.3   Root Solvers

There is a canonical way to write a rootfinder in C++, as a class with a virtual member which evaluates the objective function:

```
class RootSearch_     // deprecated interface
{
    // data representing state of the root search
public:
5    virtual ~RootSearch_();
    virtual double F(double x) const = 0;
    double Solve(double x_0, double target, double tol);
    // more Solve functions or fine-grained initializers
// ...
10 };
```

Rootfinders are written this way because the C++ virtual function replaces the C callback, which in turn replaced the Fortran 77 call of `func`. Besides this legacy, the method has little to recommend it. The derived class implementing F is a pointless blemish, and separates the evaluation code from the calling code so that the flow of control is diverted through three different classes.

The problem is that the interface is inside out: rather than *providing* a service, the rootfinder *demands* one (*i.e.*, an implementation of F). To correct this, we write instead

```
 ───────────────────── Root1.h ─────────────────────
class Rootfinder_
{
public:
   virtual ~Rootfinder_();
   virtual double NextX() = 0;
   virtual void PutY(double y) = 0;
   virtual double BracketWidth() const = 0;
};
```

A root search algorithm, like Ridders's or Brent's method, will be implemented in a concrete derived **Rootfinder_**. These methods allow the narrowing of a bracketing interval known to contain the root. The local variables of **Solve** in the callback-based implementation, which represent the state of the root search, become member data of the rootfinder:

```
 ───────────────────── Brent.h ─────────────────────
class BracketedBrent_ : public Rootfinder_
{
   pair<double, double> a_, b_, c_;
   double d_, e_, min1_, min2_;
   double p_, q_, r_, s_, tol1_, xm_;
   const double tol_;

   friend class Brent_;
public:
   BracketedBrent_
       (const pair<double, double>& low,
        const pair<double, double>& high,
        double tolerance);

   double NextX();
   void PutY(double y);
   double BracketWidth() const;
};
```

The above is based on the routine **zbrent** from *Numerical Recipes*; we have combined the two variables **a** and **fa**, which describe an inverse quadratic interpolation point, into the pair a_, and likewise for b_ and c_.

Software engineers will at this point wish to insert some checking code to ensure that PutY is never called twice in succession without an intervening NextX, and that NextX is idempotent (or likewise is never called twice). This is not difficult but also not valuable; the calling code is sufficiently simple that such errors just do not occur.

This rootfinder is still demanding a service from its client: it insists that the root be bracketed before it will do its work. To remedy this, we support an expanding "hunt" for a bracketing interval based on a single initial point:

```
──────────────── Brent.h ────────────────
class Brent_ : public Rootfinder_
{
    BracketedBrent_ engine_;
    // state for a non-bracketed rootfinder
5   enum { INITIALIZE, HUNT, BRACKETED } state_;
    bool increasing_;   // guide direction of hunt
    double stepSize_, trialX_;
    pair<double, double> knownPoint_;
public:
10  Brent_
        (double guess, double tolerance = DA::EPSILON,
        double step_size = 0.0);
    double NextX();
    void PutY(double y);
15  double BracketWidth() const;
};
```

The extra INITIALIZE state lets us create the rootfinder with just an x-value, moving all the function evaluations inside the root search loop. When the hunt has succeeded in bracketing a root, we use our friend status to initialize the Brent engine (during our PutY).[4] This lets us hold the engine data on the stack, saving a memory allocation.

Note that we favor the more complex Brent's method, rather than Ridders's. In our experience the latter takes on average 15-20% more function evaluations; thus the one-time investment in coding Brent's method results in a noticeable efficiency gain across many applications.

Now calling code becomes simple and transparent. We can make it still simpler by creating a convergence-checking utility:

---

[4]Those allergic to friends can achieve almost the same effect by using placement new to construct the engine.

```
                         ————— Root1.h —————
struct Converged_
{
    double xtol_, ftol_;
    Converged_(double x, double f) : xtol_(x), ftol_(f) {}
    bool operator()(Rootfinder_& t, double e) const
    {
        t.PutY(e);
        return fabs(e) < ftol_ || t.BracketWidth() < xtol_;
    }
};
```

For example, a bootstrapped yield curve fitter might contain this inner
loop:

```
Brent_ task(guess);
for (int i = 0; ; ++i)
{
    Require(_env, i < ITERATION_LIMIT,
            "Exhausted iterations in rootfind");
    values_->Back() = task.NextX();   // set our rate
    const double rate = target.ImpliedRate(*this);
    if (CONVERGED(task, rate - target_rate))
        break;
}
```

7.4   Underdetermined Search

But why should we use a bootstrapped yield curve fitter? Even the simplest
yield curve has thousands of degrees of freedom – e.g., the one-day forward
rates for each business day – and at most a few dozen market constraints.
The customary bootstrapping approach is a Procrustean bed upon which
"extra" degrees of freedom are lopped away until a solution is uniquely defined. But this truncation process is itself largely arbitrary, so the resulting
curve is a mass of interpolation artifacts.

Unfortunately, not only has this lesson not been learned, but the same
flawed methods have been propagated to other settings such as volatility
calibration.

Consider the following problem of calibration, with yield curve fitting
as a concrete example. We have n equality constraints to be matched, and
$M \gg n$ commensurable degrees of freedom, like the forward rates over

short non-overlapping intervals.[5] We write the constraints as $\vec{f}(\vec{x}) = 0$ where $\vec{x}$ is an M-dimensional vector representing a point in the parameter space. In the non-degenerate case, there is an $(M - n)$-dimensional manifold of solutions. Algorithms such as bootstrapping represent a recipe for picking one point on this manifold (by restricting the solution search to an n-dimensional subspace); these recipes are clearly optimized for ease of implementation, not for the quality of the output.

To pick a particular solution less arbitrarily, we must define a figure of merit – a criterion for comparing the desirability of two competing solutions. This is a general optimization problem subject to nonlinear constraints, which will be computationally impractical for even medium-sized problems. But one widely studied subclass of problems can be solved quite efficiently: the *quadratic programming* problems in which the constraints are linear and the figure of merit is a positive definite quadratic form.[6] While this problem is not quite the one we face, it is worth understanding its solution in detail.

A quadratic form of $\vec{x}$ can be written as $Q \equiv \vec{x}^T W \vec{x} + \vec{B} \cdot \vec{x} + C$; or equivalently, since W is positive definite and therefore invertible and the constant factor C does not affect the solution, as $Q = (\vec{x} - \vec{x_0})^T W (\vec{x} - \vec{x_0})$ where $\vec{x_0}$ is some point in the parameter space. While $\vec{x} = \vec{x_0}$ may not satisfy the equality constraints, it is assuredly the minimizer of Q. A positive definite matrix, like W, defines a metric on the parameter space; thus Q is the squared distance (under this metric) from the ideal solution $\vec{x_0}$. In short, quadratic programming (QP) is a search for the *closest* solution to an underdetermined problem.

The constraint function f will obviously be nonlinear in any problem of interest: even in yield curve fitting, a swap rate is not linear in any forward rate. However, a large and relevant class of problems is nearly linear, and can be well approximated by linearizing in the neighborhood of a solution. Thus the solution to the QP problem, using a linearized approximation $\vec{f} \simeq \vec{f}(\vec{x_0}) + J(\vec{x} - \vec{x_0})$, suggests an improved parameter point

$$\vec{x_1} \equiv \vec{x_0} - W^{-1} J^T (J W^{-1} J^T)^{-1} \vec{f}(\vec{x_0})$$

which, of all solutions to the linearized constraints, is closest to $\vec{x_0}$. The step from $\vec{x_0}$ to $\vec{x_1}$ is the *QP step*, which has the same role in underdetermined

---

[5]An example of incommensurable degrees of freedom is the simultaneous calibration of the various parameters of the Heston model.

[6]By convention, the problem is cast as constrained minimization rather than maximization; thus we should call this a "figure of demerit."

search that the Newton step has in the more familiar fully-specified search. A sequence of QP steps leads us to a solution which, while not exactly the closest to $\vec{x}_0$, is a very good working approximation.[7]

7.4.1   *Function and Jacobian*

The Jacobian computation is typically the most time-consuming part of this search, especially when we must resort to finite differencing. There are two ways we might optimize the computation of J: by having a fully analytic result, or by having a rapidly computable approximate objective function $\tilde{f} \simeq \vec{f}$ and computing its Jacobian by finite differencing. Obviously the first is preferable, but not always feasible.

Whenever possible, we use Broyden's update of the Jacobian rather than compute it anew:

$$ J \simeq \tilde{J} \equiv J + \frac{(\delta\vec{f} - J\delta\vec{x}) \otimes \delta\vec{x}}{||\delta x||^2} $$

where $\delta\vec{x} \equiv \vec{x}_1 - \vec{x}_0$ and $\delta\vec{f} \equiv \vec{f}(\vec{x}_1) - \vec{f}(\vec{x}_0)$ — exactly as in the fully-specified case.[8] We must monitor the approximate Jacobians to ensure that they are not too inaccurate, by comparing the expected progress toward the root with that actually obtained.

This dictates the interface of the class we will use to represent a function to the rootfinder.[9] We will enclose this within **namespace Underdetermined** to keep class names brief.

```
                        ─ Underdetermined.h ─
   class Function_
   {
       virtual double BumpSize() const;
       virtual void FFast(const Vector_<>& x,
 5         Vector_<>* f) const {*f = F(x);}
   public:
       virtual ~Function_();
       virtual Vector_<> F(const Vector_<>& x) const = 0;
       virtual void Gradient(const Vector_<>& x,
10         const Vector_<>& f, Matrix_<double>* j) const;
   };
```

---

[7]This solution, once found, can in principle be "polished" to move closer to $\vec{x}_0$ without altering $\vec{f}$. In practice this has little value.

[8]It is tempting to try Broyden's bad update here, since it has the property of seeking minimal change to the solution. I have not attempted this.

[9]Yes, I just criticized this approach for one-dimensional searches. However, the same gains in simplicity and transparency are not available here.

The default implementation of `Gradient` uses finite differencing of `FFast` — so we may override either of those functions, but generally not both at once. If neither is overridden, the method reduces to finite differencing of f. The base class supplies a default `BumpSize` for any necessary finite differencing.

Usually the weight matrix W will be quite sparse, so the dominant storage requirement is $O(Mn)$ for the Jacobian. If we need to solve a truly large problem – e.g., time-dependent path reweighting with $M \gtrsim 10^5$ – then keeping the entire Jacobian in an unstructured dense matrix is excessively memory-intensive. Thus we need to abstract the Jacobian:

```
———————————————— Underdetermined.h ————————————————
class Jacobian_
{
public:
    virtual ~Jacobian_();

    virtual int Rows() const = 0;
    virtual int Columns() const = 0;

    virtual void DivideRows(const Vector_<>& tol) = 0;
    virtual Vector_<> MultiplyRight(const Vector_<>& t)
        const = 0;
    virtual void QForm
        (const Sparse::Square_& w,
         SquareMatrix_<>* form) const = 0;
    virtual void SecantUpdate
        (const Vector_<>& dx, const Vector_<>& df) = 0;
};
```

In many cases the Jacobian can be implemented with $O(M)$ storage, like W, so the total memory requirement is $O(M + n^2)$. Such a Jacobian obviously must be computed within the `Function_`, so we add to `Function_` an additional member

```
———————————————— Underdetermined.h ————————————————
    virtual Jacobian_* Gradient
        (const Vector_<>& x,
         const Vector_<>& f)
    const {return 0;}
```

The search engine will call this function, and then if it returns 0 will call the dense-matrix implementation of `Gradient`.

7.4.2 *Weights and Smoothing*

We have talked about finding the closest solution to some initial guess, under the metric induced by W, but have begged the question of what "closest" means. It turns out that control of W is the tool we need to obtain smooth solutions.

For concreteness, consider fitting an ordered sequence of terms, like forward rates along a yield curve. If we take

$$s^T W s = \sum_i s_i^2 + \lambda \sum_i (s_i - s_{i-1})^2,$$

then W penalizes any change (the first sum) and additionally penalizes nonsmooth changes (the second term). This forms a tridiagonal matrix W with a smoothing parameter which the user can supply. As $\lambda \to 0$, we are simply asking for the smallest aggregate change in the L_2 measure; as $\lambda \to \infty$, the smoothest (smallest aggregate change to differences). The well-known trade-off between smoothness and locality is here made quantitative and explicit.

Piecewise constant forward curves – for rates, spreads, vols, and so on – are an important special case, but we cannot assume that the pieces will be of equal size. In general $\vec{x}$ describes a curve $g(t)$ such that $g(t) = x_i$ in the interval $[T_{i-1}, T_i)$.[10] Then the diagonal part of W corresponds to $\int |g|^2$, while the off-diagonal part corresponds to $\int |g'|^2$. While the latter is not defined for discontinuous functions, we can achieve the desired effect by computing it for a linear interpolant connecting the midpoints of the intervals; we obtain

$$s^T W s = \sum_i (T_i - T_{t-1}) s_i^2 + \lambda \sum_i \frac{(s_i - s_{i-1})^2}{T_i - T_{i-2}}$$

where λ now has units of t^2. Thus the user-input parameter is a *smoothing timescale* $\tau_s \equiv \sqrt{\lambda}$.

This is implemented (again, in `namespace Underdetermined`) as

```
──────────────── UnderterminedUtils.cpp ────────────────
Sparse::Tridiagonal_* WeightsPWC
    (const Vector_<Time>& knots, double t_s)
{
    auto_ptr<Sparse::Tridiagonal_> retval
        (new Sparse::Tridiagonal_(knots.size()));
    SelfCouplePWC(retval.get(), knots, t_s, 0);
```

---

[10]In a well-parametrized models only integrals of g will be observable, so the exact bounding of the interval need not concern us.

```
    return retval.release();
}
```

This forwards to a more general coupling function, supplying an additional argument which is an offset within the matrix being formed. Thus we are prepared for the task of simultaneously fitting two or more curves; for example, in calibrating a Vasicek model (Sec. 13.1). We will rely heavily on a utility function in **namespace Sparse**,

```
──────────── SparseUtils.h ────────────
inline void AddCoupling
    (Square_* dst, int i, int j, double amount)
{
    dst->Add(i, i, amount);
5   dst->Add(i, j, -amount);
    dst->Add(j, i, -amount);
    dst->Add(j, j, amount);
}
```

This reflects how we actually form sparse matrices, and saves us the tedious job of ensuring their symmetry.

7.4.3 *Monitoring Progress*

To make sure that we are progressing towards a root, we extend the technique of *backtracking linesearch* used in multi-dimensional root search. Given two points, say $\vec{x}_0$ and $\vec{x}_1$, and their corresponding function values, we define $\tilde{f}(k) \equiv k\vec{f}(\vec{x}_0) + (1-k)\vec{f}(\vec{x}_1)$. Then we compute the *overshoot fraction* $k_{\min}$, the value of k which minimizes $|\tilde{f}|$.

We may choose to backtrack, using some *backtracking fraction* $\bar{k}$: that is, replace x_1 with $(1-\bar{k})x_1 + \bar{k}x_0$, evaluate $\vec{f}$ again at this new point, and restart the backtracking process.

The rootfind succeeds when $|f_i| < \epsilon_i$ for all i, for a user-supplied tolerance $\vec{\epsilon}$. It fails if we exhaust pre-determined limits on the allowed number of function evaluations or gradient evaluations, or if we cannot decrease $|f|^2$ even after frantically backtracking – *i.e.*, if the Jacobian freshly evaluated by **Gradient** does not point in a descent direction.

Thus the search routine has many control parameters, which we encapsulate in a structure (again in **namespace Underdetermined**) with sensible values provided by a default constructor.[11] In fact, we do not write this class, but describe it using mark-up:

```
——————————— UnderdeterminedControls.settings.if ———————————
  INTEGER maxEvaluations COND{@me>0}
  ' Give up after this many point evaluations.
  INTEGER maxRestarts COND{@me>0)
  ' Give up after this many gradient calculations.
5 INTEGER (DEFAULT 5) maxBacktrackTries
  ' Give up if no descent step is found by then.
  NUMBER (DEFAULT 0.15) restartTolerance
  ' Restart when k_min is above this
  ...
```

This defines a `struct` which can be formed from a user-input `Dictionary_`, or formed internally and altered as needed. Our interface parser will create a corresponding `struct`, constructible from and convertible to a `Dictionary_`; and will invoke the required translation code at the public interface.

```
——————————————————— Underdetermined.h ———————————————————
  Vector_<> Find
     (const Function_& func,
      const Vector_<>& guess,
      const Vector_<>& tol,
5     const Sparse::Square_& w,
      const Controls_& controls,
      Matrix_<double>* eff_j_inv = 0);
```

The last argument, which asks for the final value of $W^{-1}J^T(JW^{-1}J^T)^{-1}$, is used to prepare for repeated calibrations – *e.g.*, when computing vega risk by bumping each calibration instrument price.

7.5 Quadrature

Quadrature, or numerical integration, is the mirror image of root search; as with root search, we write an integrator interface so as to avoid having to wrap the integrand in its own class. Two additional complications arise: the integrand may be vector-valued, and the integration may be adaptive. We also might wish to reuse an integrator (whereas a rootfinder is always used and then thrown away).

---

[11] We can also change the search engine to extend too-short steps (those with $k_{min} < 0$); this will of course be controlled by still more parameters.

```
──────────── Quadrature.h ────────────
   template<class T_> class Quad1D_
   {
   public:
      virtual ~Quad1D_();
5     virtual double GetX() = 0;
      virtual void PutY(const T_& y) = 0;
      virtual bool IsComplete() const = 0;
      virtual T_ Result() const = 0;
      virtual void Restart() = 0;
10 };
```

Clearly `Result` should be called only if `IsComplete` returns `true`; we
will check this with an assertion inside implementations of `Result`. The
`Restart` function restores the object to its newly-constructed state for
reuse.

7.5.1 *Gaussian Quadrature*

A fixed integrator (e g , Gauss-Hermite) sets the abcissas without any feed-
back from the function; it is just a weighted sum of evaluations. We represent this in a derived class:

```
──────────── Quadrature.h ────────────
   template<class T_> class Quad1DFixed_ : public Quad1D_<T_>
   {
      size_t i_;    // because STL sizes are size_t
      T_ sum_, initial_;
5  protected:
      Vector_<> x_, w_;
      Quad1DFixed_(int size, const T_& initial)
         : x_(size), w_(size), i_(0),
           initial_(initial), sum_(initial)
10    {   }
   public:
      double GetX() {assert(!IsComplete()); return x_[i_];}
      void PutY(const T_& y)
      {
15       assert(!IsComplete());
         Quadrature::Increment(&sum_, y, w_[i_++]);
      }
      bool IsComplete() const {return i_ == x_.size();}
      T_ Result() const {assert(IsComplete()); return sum_;}
20    void Restart() {i_ = 0; sum_ = initial_;}
      // allow query
```

```
    const Vector_<>& Abcissa() const {return x_;}
    const Vector_<>& Weight() const {return w_;}
};
```

A specific integrator, such as Gauss-Legendre, is now implemented as a subclass of this function whose constructor calls a (non-template) function to populate the `abcissa_` and `weight_` vectors. Thus the complexity of Gaussian quadrature is encapsulated in these few functions, while the scaffolding of integration over predetermined abcissae is collected in the template classes we have just displayed.

This implementation is supported by a function to encapsulate `s += w * y`:

———————————— *Quadrature.h* ————————————

```
namespace Quadrature
{
    template<class T_> void Increment
        (T_* dst, const T_& inc, double w)
    {
        T_ z(y); z *= weight_[i_++]; sum_ += z;
    }
    template<> void Increment   // specialization
        (Vector_<>* dst, const Vector_<>& inc, double w)
    {
        Transform(dst, inc, AddMultiple(w));
    }
}
```

The template specialization avoids an unnecessary vector copy when the integrand is itself a `Vector_<>`.

It is worth saying a few words about Gauss-Hermite integration in particular. The use of this in practice is for computing the expectation of a function of a normal deviate; but the conventions of Gauss-Hermite (with a weight $\exp\{-x^2\}$) are off by constant factors in both x and w from those of the normal distribution (with $\phi(x) = \exp\{-x^2/2\}/\sqrt{2\pi}$). Rather than make repeated *ad hoc* corrections for this, it is better to have an integrator class take care of it once and for all. Similarly, we will wrap Gauss-Laguerre integration in `class GammaExpectation_`, which comports with its most common use.

———————————— *QuadratureGaussian.h* ————————————

```
template<class T_ = double> class NormalExpectation_
    : public Quad1DFixed_<T_>
{
```

```
public:
    NormalExpectation_(int n, const T_& initial = 0.0)
        : Quad1DFixed_<T_>(n, initial)
    {
        Quadrature::NCDFGaussHermiteWeights(&x_, &w_);
    }
};
```

We supply `0.0`, not a more general default, because there is no good default value for vector integrands: we must start with a vector of the appropriate size. Our code gives a compilation error if we attempt

```
NormalExpectation_<Vector_<> > integrator(10);
```

but allows the obvious default for scalar integrands.

7.5.2 *Adaptive Quadrature*

Adaptive integrators, such as Romberg's, are intrinsically unstable: for two near-identical inputs, where we would want to see near-identical results (e.g., when computing a parameter sensitivity by finite differencing), the signal can be swamped by a change in the integrator's strategy. The signature of this instability is infrequent and seemingly random blowouts in computed risk figures. This can be avoided with sufficient machinery (see Secs. 5.6, 10.9.1, and 14.4); but for most applications we will favor fixed integrators for their greater speed and stability, despite the lack of guaranteed convergence.

7.6 Distributions

Another common building block is a one-dimensional (likely risk-neutral) probability distribution. The lognormal distribution is the canonical example. We begin with an abstract base class, whose specification reveals our intentions for its use:

```
———————————— Distribution.h ————————————
class Distribution_
{
public:
    virtual ~Distribution_();
    virtual double Forward() const = 0;
    virtual double OptionPrice
```

```
                (double strike, const OptionType_& type)
            const = 0;
            // support calibration of whatever vol-like parameter
10          virtual double& Vol() = 0;     // whatever it means
            virtual const double& Vol() const = 0;
            virtual double VolVega
                (double strike, const OptionType_& type)
            const = 0;
15          // support hedge computation
            virtual Vector_<String_> ParameterNames() const = 0;
            virtual map<String_, double> ParameterDerivatives
                (double strike, const OptionType_& type,
                 const Vector_<String_>& to_report)
20          const = 0;
    };
```

We assume that every distribution has a unique vol-like parameter: log-normal vol, normal vol, the SABR α, and so on. The oddly named VolVega is a *proportional vega*, the product of vol and vega (or the derivative of price with respect to log of vol). This turns out to be much more useful than vega alone for calibration; it is very similar across different distributions with different meanings of "vol" (*e.g.*, normal and lognormal distributions). Also, it has no units of time, so it is the same for annualized or deannualized vols; this property is useful when implementing stochastic-time distributions (such as gamma variance) which must themselves report some form of vega.

The last two functions will be used to support "market models" for swaptions, equities or FX – these are distribution-based pricing models with no specification of the dynamical process from which that distribution arises, and thus are valid only for simple European options.[12] Armed with this functionality, we can build the framework of such a model without committing to any particular parametrization of the distribution; see Sec. 14.6.

7.6.1 *Implied Vol*

One recurrent problem is the need to compute reliable implied vols for deep in- or out-of-the-money options which have extremely little value; the option price may be indistinguishable from the intrinsic value. A straightforward

---

[12]Jamshidian *et al.* have attempted to back out dynamics from swap market models, but I am unaware of any practical tools to emerge from this.

implied vol computation will thus produce numbers which are correct, in the sense that they reproduce the option price to good accuracy, but are less than intuitive. We can choose how to spend our time: in explaining to users why this does not matter, or in implementing a more robust implied vol.

Given a distribution (rather than just a single option price), we can accomplish the latter goal by stepping outward from the forward to the actual strike, using the implied volatility from the previous iteration (at first, the at-the-money vol) as the initial guess each time. To accomplish this, we write

```
────────────────── DistributionUtils.cpp ──────────────────
double Distribution::BlackIV
    (const Distribution_& model, double time,
    double strike, double guess, int n_steps)
{
    const double f = model.Forward();
    const OptionType_ type = strike > f
            ? OptionType_::CALL
            : OptionType_::PUT;
    if (n_steps > 1)
    {
        const double fMid - strike > f
                ? strike * pow(f / strike, 1.0 / n_steps)
                : strike + (f - strike) / n_steps;
        guess = BlackIV(model, time, fMid, guess, n_steps-1);
    }
    return BlackIV
            (time, f, strike, type,
            model.OptionPrice(strike, type), guess);
}
```

where the last call is to the usual form of BlackIv. This is a non-tail recursion, and it is not worth the effort of unwinding it into an iteration; usually half a dozen steps gives an answer of very good quality.

7.7 Baskets

Basket options are an important class of equity derivatives; they are also a crucial tool for swaption pricing in Libor-based interest rate models (Goldman-Pugachevsky, BGM or "string" models) where a swap is approximated as a basket of FRAs (forward rate agreements) or futures. Both

Derivatives Algorithms

problem descriptions are similar: we have a collection (the "basket") of assets with similar dynamics, and an option on a weighted sum of their values, so we must approximate $E\left[(B - K)^+\right]$ where $B \equiv \sum_i w_i S_i$.

Clearly the weights are not central to the problem, since we can absorb the scale factor into S_i. Once the dynamics of the S_i are specified, we can take one of two approaches: *whole-basket* methods which approximate the distribution of the total basket value, or *effective-strike* methods which use the strike K to extract relevant statistics for each S_i. In our experience, effective-strike methods are faster and better behaved in the extreme tails of the distribution, though they are trickier to implement.

7.7.1 *Whole-Basket Moment Matching*

If the S_i are multivariate normal, then B is also normal and the problem is trivial. If they are multivariate lognormal, then B is not lognormal, but we can approximate it by a shifted lognormal. For this purpose we need the first three moments of B. These can be accumulated in a simple loop:

```cpp
                              ─── Basket.cpp ───
double m1 = 0.0;
double m2 = 0.0, m22 = 0.0;
double m3 = 0.0, m32 = 0.0, m33 = 0.0;
for (int i = 0; i < n; ++i)
{
    if (IsZero(weights[i]))
        continue;
    const double si = weights[i] * forwards[i];
    m1 += si;
    m2 += Square(si) * ec(i, i);
    m3 += Cube(si * ec(i, i));
    for (int j = 0; j < i; ++j)
    {
        if (IsZero(weights[j]))
            continue;
        const double sj = weights[j] * forwards[j];
        m22 += si * sj * ec(j, i);
        m32 += si * sj * Square(ec(j, i))
                * (si * ec(i, i) + sj * ec(j, j));
        for (int k = j + 1; k < i; ++k)
        {
            const double sk = weights[k] * forwards[k];
            m33 += si * sj * sk
                    * ec(j, i) * ec(k, i) * ec(k, j);
        }
    }
```

```
      }
}
m2 += 2.0 * m22;
m3 += 3.0 * m32 + 6.0 * m33;
```

Here `ec` is the matrix of exponentiated covariances:

$$\texttt{ec(i,j)} = \exp\Big\{\text{Cov}[\log S_i, \log S_j]\Big\}.$$

Creating this matrix saves $O(n^3)$ calls to `exp`, so it is a very worthwhile optimization. The choice of the loop conditions for j and k – so that $j < k < i$, rather than the more usual $k < j < i$ – is due to the routine's importance for swaption pricing: we can have several zero weights at the front (before expiry) and the back (after swap maturity) of the basket, and we can omit these terms without having to put an extra test in the innermost loop.

In practice this clean inner loop will be obscured by the need to accumulate optional delta and vega outputs. We will translate the moments to a lognormal part (forward and vol) and a shift, and use these in pricing; sensitivity to these parameters can then be mapped back to sensitivity to the moments themselves, and thence to the input parameters, by repeated application of the chain rule. Thus the relevant intermediate quantity is the sensitivity of each moment to each of the forward prices $E[S_i]$ or covariances; we will accumulate these within the loop and combine them at the end, just as we have done for m_2 and m_3.

Fitting a shifted lognormal distribution to the computed moments requires the solution of a cubic equation. Let $y \equiv \exp\{\sigma^2\}$ where σ is the volatility of the lognormal part; then

$$(y-1)(y+2)^2 = \frac{\left[\frac{m_3 - m_1^3}{m_2\, m_1^3} - 3m_1\right]^2}{m_2 - m_1^2}.$$

If we are computing sensitivities, the root of the cubic appears again in the chain-rule application; thus it must be provided as an optional output.

— *Basket.cpp* —

```
void XThreeMomentFit
    (_ENV, double m1, double m2, double m3,
     double* ln_part, double* shift, double* vol,
     double* root);
```

7.7.2   Taylor Expansion of Projected Vols

If we define
$$B \equiv \sum w_i S_i, \qquad Q_i \equiv \sum_j \sigma_{ij} w_j S_j, \quad \text{and} \quad V \equiv Q \cdot S = (wS)^T \sigma(wS),$$
then we can compute the elasticity of the basket price B:
$$\beta = \frac{1}{2} \frac{dV \wedge dB}{dB \wedge dB} \left(\frac{V}{B} \right)^{-1} = \frac{(\sum_j Q_j^2 w_j S_j)(\sum_j w_j S_j)}{(\sum_j Q_j w_j S_j)^2}.$$
We will always have $\beta \geq 1$ by Cauchy's inequality. Thus we can approximate the basket as a shifted lognormal distribution, where the shift is $\Delta \equiv B(1 - \beta^{-1})$. The volatility of the lognormal can be set to $\sqrt{V}/(B - \Delta)$, which is correct at time 0, or by matching the second moment of the total basket price. This computation takes only $O(n^2)$ time, and is competitive with whole-basket methods in accuracy.

7.7.3   Midpoint Variance

A better method for matching the total basket variance is to use the strike; or, more specifically, the characteristics of the mean path from spot to strike. We are interested in the basket's volatility when $B \simeq (E[B] + K)/2$; analysis of this midpoint captures most of the benefits of path integration.

Define the relative strike $M = K - E[B]$, and for each i compute the "strikelet"
$$\tilde{k}_i \equiv E\left[S_i | B = K \right] = E[S_i] + \frac{MQ_i}{w_i \sum_j Q_j} + O(M^2).$$
We can convert each lognormal volatility to a normal volatility by taking the instantaneous conversion at the midpoint, $\sigma_i^{(N)} = \sigma_i^{(ln)}(E[S_i] + \tilde{k}_i)/2$; or by inferring a normal vol from the value of the option struck at $\tilde{k}$. The latter approach is more computationally intensive, but more robust for highly inhomogenous baskets.

7.8   Random and Quasi-Random Numbers

7.8.1   Random Deviate Streams

A random[13] generator should encapsulate both the production of uniform deviates and their transformation into normal deviates. For most dynamical

---

[13]These are deterministic and therefore, strictly speaking, only "pseudo-random." We nonetheless prefer the briefer nomenclature.

models, we will draw batches of normal deviates; however, uniform deviates are more convenient for simulation of discrete events (events of default, and the jump processes of several models). We will support this with the following interface:

```
───────────────────────── Random.h ─────────────────────────
class Random_
{
public:
    virtual ~Random_();
    virtual double NextUniform() = 0;
    virtual void FillUniform(Vector_<>* deviates) = 0;
    virtual void FillNormal(Vector_<>* deviates) = 0;
    virtual Random_* Branch(int i_child = 0) const = 0;
};
```

The first three functions draw random numbers, changing the class state in the process. `Branch` creates an independent random stream; see Sec. 10.9.

A sensitivity is computed from the difference of the price computed with the input model (the *base run*) and the same price computed with a slightly different *bumped model* (the *bumped run*). We want to exclude any source of variation in price except for the difference in models; thus, for instance, we want to use the same deviates in the same order.

We might think to store all the random deviates used in the base run, and simply apply them again during the bumped run. This approach has two related weaknesses. First, the extra memory requirement reduces the maximum number of paths we can run on a given machine, which for long-running trades in very complex models is already an unpleasant constraint. Second, the additional memory requirement increases the overhead of the computation, replacing productive CPU usage with cache faults, so it is probably faster to regenerate the deviates than to save them.

Whether random deviates are stored or re-created, the pattern in which they are used – the mapping from the random stream to financial observables – must be unchanged. We accomplish this by using the same number of deviates at each step on each path, so that the state of the random stream will be the same at the corresponding point in both runs.

7.8.2 *Generator Implementation*

Though `FillUniform` is declared pure virtual, we will supply the obvious implementation:

```
————————————— Random.cpp —————————————
void Random_::FillUniform(Vector_<>* devs)
{
    for (auto pd = devs->begin(); pd != devs->end(); ++pd)
        *pd = NextUniform();
5 }
```

This function can be called explicitly by a derived class, using the syntax `Random_::FillUniform(deviates)`. It is not supplied as a full-fledged default because of its inefficiency: it requires a virtual function call for each deviate. A production generator should avoid this.

There are several good choices for generation of uniform deviates. Our favorite methods are

(1) The "Mersenne twister" of Matsumoto and Nishimura, our first choice.
(2) Numerical Recipes `Ran2`, a reliable standard.
(3) A pair of generators like Knuth's `IRN55`, with the second shuffling the outputs of the first.

All these are readily available online.

The option to switch generators at runtime seems to have minimal value, so we prefer to settle on one of these generators and use it for all pricing. Thus we create random generators through the interface

```
————————————— Random.h —————————————
namespace Random
{
    Random_* New(int seed);
}
```

There is no need for `seed` to be negative, since we have abandoned the (very restrictive) trick of using it to smuggle the random generator's state. Acclimation to a world of positive seed values is quick and painless.

7.8.3 Transforms

Another question is how to effect the transformation of uniform to normal deviates. Again, there are several plausible candidates:

(1) Directly through the inverse cumulative normal density.
(2) The polar method of Box and Mueller, most widely used.
(3) The odd-even method of Von Neumann and Forsyth.
(4) The rectangle-wedge-tail method of Marsaglia, the fastest available.

Knuth (in *Art of Computer Programming*, volume 2) gives an implementation of Marsaglia's method but leaves computation of its many constant parameters as an exercise for the reader. In addition, Knuth's routine partitions the distribution into 32 segments; by using a finer partition, we increase the size of the tables of constants but decrease the frequency of calls to the math library. For example, with partition into 64 segments we have the static data

```
                    ──── Random.cpp ────
   static const double RWT_D[31] = {0.0, 0.0, 0.0, 0.0,
      0.0, 0.0, 0.0, 0.0, 0.0, 0.0, 0.0, 0.0, 0.0, 0.0,
      0.0, 0.0, 0.5050335007, 0.7729568318, 0.8764243173,
      0.9392112429, 0.9860868156, 0.9951545013,
5     0.9867480142, 0.9792113586, 0.9722739162,
      0.9657523400, 0.9595309729, 0.9535340961,
      0.9477102649, 0.9420234020, 0.9364475249};
   static const double RWT_E[31] = {0.0, 0.0, 0.0, 0.0,
      0.0, 0.0, 0.0, 0.0, 0.0, 0.0, 0.0, 0.0, 0.0, 0.0,
10    0.0, 0.0, 25.0, 12.5, 8.3333333333, 6.25, 5.0,
      4.0637731069, 3.3677961409, 2.8582959135,
      2.4694553648, 2.1631696640, 1.9158499112,
      1.7121118654, 1.5414940825, 1.3966346593,
      1.2722024279};
15 static const double RWT_P[32] = {0.0, 0.8487410443,
      0.9699899979, 0.8550231215, 0.9942754672,
      0.9951625205, 0.9327422986, 0.9233994671,
      0.7273661575, 1.0, 0.6910843714, 0.4540747884,
      0.2866499878, 0.1738620062, 0.1013177803,
20    0.0567276597, 0.0672735098, 0.1605108070,
      0.2355403454, 0.2854029087, 0.3075794736,
      0.3038922909, 0.2795217494, 0.2414883115,
      0.1970555059, 0.1524486512, 0.1121116518,
      0.0785256947, 0.0524616474, 0.0334682128,
25    0.0204066391, 0.0063979023};
   static const double RWT_Q[16] = {0.0, 0.2356431344,
      0.2061876931, 0.2339118030, 0.2011514983,
      0.2009721989, 0.2144214970, 0.2165909849,
      0.2749646762, 0.2, 0.2894002647, 0.4404560771,
30    0.6977150132, 1.1503375831, 1.9739871865,
      3.5256169769};
   static const double RWT_S[17] = {0.0, 0.0, 0.2, 0.4,
      0.6, 0.8, 1.0, 1.2, 1.4, 1.6, 1.8, 2.0, 2.2, 2.4,
      2.6, 2.8, 3.0};
35 static const double RWT_Y[32] = {0.0, -0.922235758,
      -5.864444728, -0.579530323, -33.13734925,
      -39.54384419, -2.573637156, -1.610947892,
```

```
         0.666415357, DA::NAN, 0.352574032, -0.166350547,
         0.919632724, 0.357909694, -0.022548077, 0.187972157,
40       0.585574869, 0.961759887, -0.061622701, 0.120122007,
         1.311158187, 0.312688141, 1.12240661, 0.536325751,
         0.75091678, 0.564026097, 0.174746453, 0.382956509,
         -0.01107325, 0.393074576, 0.195833651, 0.781086317}};
    static const double RWT_Z[32] = {0.2, 1.3222357584,
45       6.6644447279, 1.3795303233, 34.9373492509,
         41.3438441882, 2.9736371558, 2.6109478918,
         0.7335846430, DA::NAN, 0.6474259683, 0.3663505472,
         0.2803672763, 0.2420903063, 0.2225480772,
         0.2120278433, 0.2144251312, 0.2382401128,
50       0.2616227015, 0.2798779934, 0.2888418127,
         0.2873118590, 0.2775933900, 0.2636742492,
         0.2490832199, 0.2359739033, 0.2252535472,
         0.2170434909, 0.2110732504, 0.2069254241,
         0.2041663490, 0.2189136830}};
```

This supports the algorithm, which closely follows Knuth:

```
                       ──────── Random.cpp ────────
    template<class SRC_> FORCE_INLINE void FillNormalRWT
       (SRC_* src, Vector_<>::const_iterator dst_begin,
       Vector_<>::const_iterator dst_end)
    {
5       for (auto pn = dst_begin; pn != dst_end; ++pn)
        {
            double f = 64.0 * src->NextUniform();
            double sign = (f > 32.0 ? 1.0 : -1.0);
            int j = int(f);
10          f -= j;
            j &= 31;
            // f is uniform in [0, 1), j in {0..31}
            if (f >= RWT_P[j])    // 60.55%
                *pn = RWT_Y[j] + f * RWT_Z[j];
15          else if (j < 16)      //31.28%
                *pn = RWT_S[j] + f * RWT_Q[j];
            else if (j < 31)      // 7.90%
            {
                double u, v;
20              do        // loop c. 1.056 times
                {
                    u = src->NextUniform();
                    v = src->NextUniform();
                    if (u > v)
25                      swap(u, v);
```

```
                *pn = RWT_S[j - 15] + 0.2 * u;
                if (v <= RWT_D[j]) // In triangle
                   break;
             }   // full rejection test:  1.00%
30           while ((exp(0.5 * (Square(RWT_S[j - 14]) -
             Square(*pn))) - 1.0) * RWT_E[j] + u < v);
          }
          else    // "Supertail" case; 0.27%
          {
35           do    // loop 1.094 times
             {
                const double u = src->NextUniform();
                *pn = sqrt(9.0 - 2.0 * log(u));
             }
40           while (*pn * src->NextUniform() >= 3.0);
          }
          *pn *= sign;
       }
}
```

The `inline` directive may not be strong enough; many compilers will judge the code too complex for inlining. Thus we use `FORCE_INLINE`, a macro which will be resolved as a platform-dependent directive. This routine uses on average about 1.17 calls to `NextUniform`, and calls a math library function (`exp`, `sqrt` or `log`) once for every 60 normal deviates.

If the Marsaglia algorithm is deemed too complex, the strongest alternative is the direct method, due to its simplicity of implementation.

7.8.4 *Low-Discrepancy Sequences*

The convergence of Monte Carlo methods can in many cases be greatly improved by using numbers which, rather than emulate randomness, approach a uniform distribution over the domain of integration (which for this purpose is mapped onto a D-dimensional unit hypercube, $[0,1]^D$). Several families of such sequences are known. A few are not useful for one reason or another:

- *Halton sequences* depend on a different prime base for each dimension; if the dimension is significant, then the prime bases become large and successive paths slowly "scan" along a few low-dimensional manifolds within the unit cube.

- *Faure sequences* depend on a prime base $P \geq D$. The coverage after P^{P-1} paths is excellent; then the next P^{P-1} paths nearly duplicate the previous, forming a smaller-scale cover which is not complete until we reach P^{P^2-1} paths. This discourages small values of P, but for large P we again see the "scanning" behavior. Faure sequences are also not well adapted to binary computation.

The most useful low-discrepancy sequences are those of Sobol and of Niederreiter. They seem to perform equally well in practice; here we will focus on Sobol sequences, whose implementation is somewhat simpler.

We must always bear in mind that Sobol (or other) sequences are an intrinsically unique resource; we must ensure that a given sequence is used only once in a Monte Carlo computation. This is simple enough for simple models, but requires care when we start constructing hybrid models with multiple components, each consuming random or quasi-random numbers.

We support this resource protection by defining the functionality of a set of sequences:

```
                    ──────── QuasiRandom.h ────────
    namespace QuasiRandom
    {
        class SequenceSet_ : noncopyable
        {
        public:
            virtual ~SequenceSet_();
            virtual int Size() const = 0;
            virtual void Next(Vector_<>* dst) = 0;
            virtual SequenceSet_* TakeAway(int subsize) = 0;
        };
    }
```

Each call to `Next` populates `dst` with numbers in $(0,1)$, and advances the path index or state of the sequences as necessary. The `TakeAway` function partitions the sequence set; its input is the number of sequences to place into the returned subset, while the remaining sequences are kept in `this`. Thus `TakeAway(Size())` returns a copy of the existing sequences, emptying `this` itself in the process. To support hybrid models as described above, we will construct a large `SequenceSet_` at the top level, and call `TakeAway` to extract non-interfering subsets for each submodel.

A Sobol sequence is generated from a set of *direction numbers*, which in turn obey a recursion based on a given primitive polynomial modulo 2. We use one sequence for each polynomial: two sequences based on the same polynomial, even with different initial direction numbers, provide very poor coverage of the unit square.

Once the direction numbers are known, the generating polynomial is no longer needed. Thus our Sobol sequence set does not contain it.

```
———————————————————————————————— Sobol.h ————————————————————————————————
class SobolSet_ : public QuasiRandom::SequenceSet_
{
    Matrix_<int> directions_;   // index as [i_bit][i_seq]
    int iPath_;
5   Vector_<int> state_;

    friend struct SobolInitializer_;
    SobolSet_(int size, int i_path);
public:
10  int Size() const {return state_.size();}
    void Next(Vector_<>* dst);
    SobolSet_* TakeAway(int subsize);
};
```

We start with a path index `i_path` $\gg 1$. The reason is a particular property of Sobol sequences: the first entry in each sequence is $\frac{1}{2}$, which is unusually well centered; then $\frac{1}{4}$ and $\frac{3}{4}$ in either order; and so on. In general the first n paths have too little dispersion: the average of $N^{-1}(x_i)^2$ is roughly $1 - \log_2 n/n$. This can distort computed option prices; we avoid the problem by taking our paths – a few thousand in number – from a randomly chosen index among the billion or so possible. To accomplish this, our initializer must be able to "fast forward" to a given path. This turns out not to be difficult:

```
——————————————————————————————— Sobol.cpp ———————————————————————————————
Fill(&seq->state_, 0);
for (int jj = 0, ip = seq->iPath_; ip; ++jj, ip >>= 1)
{
    if ((ip ^ (ip >> 1)) & 1)
5   {   // the Gray code of iPath_ has a 1<<jj bit
        Transform(&seq->state_, seq->directions_.Row(jj),
            ptr_fun(Xor));
    }
}
```

This is desirable for another reason: the variation seen by the user as the seed changes should encompass all sources of numerical error in the price. If we always start from the first Sobol path, we are hiding such a source of error, so the average over many runs will converge to a false estimate of the price.

The implementation of `Next`, following *Numerical Recipes*, is quite efficient:

```
———————————————— Sobol.cpp ——————————
   double ScaleTo01(int state)
   {
       static const double MUL = 0.5 / (1L << (N_BITS - 1));
       return MUL * state;
 5 }
   int Xor(int i, int j) {return i ^ j;}

   void SobolSet_::Next(Vector_<>* dst)
   {
10     dst->resize(Size());    // usually no-op
       ++iPath_;
       assert(iPath_ != 0);    // so next loop can terminate
       int k = 0;
       for (int j = iPath_; !(j & 1); j >>= 1, ++k);
15     assert(k < directions_.Rows());
       Transform(&state_, directions_.Row(k), ptr_fun(Xor));
       Transform(state_, ScaleTo01, dst);
   }
```

Note that we store an `int` for each sequence's state, but convert to a double in $(0, 1)$ for output. The weird computation of `MUL` is designed to prevent overflowing the positive range of integers. We use `N_BITS == 30`; we could grab an extra bit by using `unsigned int`s, but it has no real value.

For large D, the computation of the first D primitive polynomials can be time-consuming. It is better to compute offline the first few thousand such polynomials and insert them directly into the code.

```
———————————————— Sobol.cpp ——————————
   static const unsigned int KNOWN_PRIMITIVE[N_KNOWN] =
       {0x0000002, 0x0000003, 0x00000005, 0x0000000A, ...
```

We could omit the 1 bit in this representation (adding special handling for the first polynomial $p(x) = x$) — if we were demented enough to foresee the need to handle more than about 50 million sequences.

7.8.5 *Spectral and Spining Methods*

Sobol (and other low-discrepancy) sequences perform best in low-dimensional problems. As the dimensionality increases, it becomes harder to obtain good convergence even for a simple payout.

This is exacerbated by embedding in our simulation code a naive concept of dimension. If D is simply defined as the number of random deviates consumed in the process of stepping to the trade's maturity, then D will rapidly grow large. But many trades are *directional*, so that the payout of the trade is largely determined by a total (time-integrated) innovation, with little dependence on other details of the path.

We can use directionality to improve convergence, if we can assign Sobol numbers directly to the random deviates most important to the trade's payout. This idea, called *reduction of effective dimension*, has been implemented using recursive Brownian bridging (pioneered by William Morokoff) or "spining" which assigns Sobol deviates to a few partial sums (developed by the author). Extending these techniques to multi-driver models requires some work.

An interesting property of the Fourier transform is that the transform of a vector of standard $N(0, 1)$ deviates is itself a vector of normal deviates. This suggests a *spectral* approach, more effective than spining and more generic than Brownian bridge, to the assignment of Sobol sequences.

For each driver, we will form a path of random deviates and then Fourier-transform it. There is one wrinkle: since FFT is only available for arrays whose size is a power of 2, we must round up the size before transforming. After transforming, we combine pairs (using $x_i = (y_j + y_{j+1})/\sqrt{2}$) to regain the desired output size.

7.9 PDE Solvers

While our most important models are high-dimensional and require Monte Carlo methods, there are still special cases where finite-difference PDE solvers are valuable. We will have no use for fully explicit "tree" solvers, which are inferior in every application.[14] A full treatment of the science of PDE solvers would require a volume far larger than this one. However, without too much trouble we can build a solver which is unconditionally stable and reasonably accurate. A good general-purpose scheme is described in Lipton's *Mathematical Methods in Foreign Exchange*.

We wish to abstract the PDE solver – a tool to implement a finite-differencing scheme – from the calling routine, which provides the terminal conditions and maybe free boundaries. Passing the terminal condition and coefficient calculators to a "generic" solving function is wrong, for the same

---

[14]See especially our discussion of forward induction, below.

reasons discussed in Sec. 7.3: it unnecessarily forces us to bundle the calculators into a class whose very existence is a design flaw. Instead, we encapsulate the scheme as an independent object, derived from an abstract base class in **namespace** PDE:

```
                          ───────── PDE.h ─────────
    class Rollback_
    {
    public:
        virtual ~Rollback_();
5
        virtual void Step
            (double dt,          // positive
             const vector<CoordinateVector_>& x_points,
             const vector<shared_ptr<Cube_> >& old_vals,
10           const ScalarCoeff_& discounting,
             const VectorCoeff_& advection,
             const MatrixCoeff_& diffusion,
             vector<shared_ptr<Cube_> >* new_vals)
        const = 0;
15  };
```

Now let us describe the lower-level interfaces we have just introduced.

7.9.1 *Cube*

A cube is a three-dimensional data structure with dense data; thus it can store the values at each point of a grid of up to three dimensions. This interface choice limits our PDE solver to at most three spatial dimensions; we do not chafe at this restriction, since PDE solvers are asymptotically inferior to Monte Carlo beyond this point.

In one dimension, it would seem better to have a **Vector_<>** than a **Cube_**; we minimize this drawback by supporting an efficient **Swap** of the contents of a vector and an $1 \times 1 \times n$ cube. This requires that, at the least, the contents of a one-dimensional "slice" must be stored in a vector; in practice we will store the cube's entire contents in a **Vector_<>**.

```
                          ───────── Cube.h ─────────
    class Cube_
    {
        Vector_<> vals_;
        Vector_<double*> hooks_;      // pointers into vals_
5       Vector_<double**> slabs_;     // pointers into hooks_
        int sizeJ_, sizeK_;           // sizeI = sizeof(slabs_)
        void SetHooks();
```

```
    public:
        Cube_() : sizeJ_(0), sizeK_(0) {}
10      Cube_(int n_i, int n_j, int n_k, double fill = 0.0);
        Cube_(const Cube_& src);        // have to SetHooks
        Cube_& operator=(const Cube_& src);

        inline const double& operator()
15          (int ii, int jj, int kk) const
        { return slabs_[ii][jj][kk]; }
        inline double& operator()(int ii, int jj, int kk)
        { return slabs_[ii][jj][kk]; }
        inline double* SliceBegin(int ii, int jj)
20      { return slabs_[ii][jj]; }
        inline const double* SliceBegin(int ii, int jj) const
        { return slabs_[ii][jj]; }
        inline const double* SliceEnd(int ii, int jj) const
        { return SliceBegin(ii, jj) + sizeK_; }

25
        inline int sizeI() const {return slabs_.size();}
        inline int sizeJ() const {return sizeJ_;}
        inline int sizeK() const {return sizeK_;}
        bool empty() const;

30
        void resize(int size_i, int size_j, int size_k);
        void fill(double value);
        void swap(Cube_* other);
        void swap(Vector_<>* other);    // need SizeI==SizeJ--1
35      Cube_& operator*=(double scale);
    };
```

7.9.2 *Coordinate Mapping*

A model communicates coefficients (e.g., the diffusion coefficient $v^2/2$ in a normal model) in its own coordinate space, the same space in which its state variables are specified. We can often improve PDE convergence by using nodes which are not equidistant in this coordinate space; the best way to do so without a great loss of efficiency is to solve the PDE on a regular grid in some transformed coordinate space. This can involve any mapping $\mathcal{R}^n \to \mathcal{R}^n$; in practice we will consider only separable mappings, so we specify a function mapping $\mathcal{R} \to \mathcal{R}$ in each spatial dimension. This means that correlations must be handled by the solver, since they cannot be rotated away in a coordinate transformation.

We also allow the mapping to be time-dependent, but only in a very restricted way: we can rescale all the y_i, at some specified set of times, by some constant factor. The mapping function $x(y)$ is not affected; time-dependent mapping functions do not bring benefits commensurate to their cost in code complexity and execution time. Our more restrictive scheme allows us to "refocus" the grid at short timescales – when the envelope of likely x-values is smaller – at minimal cost.

Our demands on the coordinate mapping are not symmetric between x and y. We need to compute not only $x(y)$ but also its first two derivatives, in order to know the PDE coefficients in y-space; but in the other direction we need only compute $y(x)$.

```
────────────────── PDE.h ──────────────────
class CoordinateMap_ : noncopyable
{
public:
    virtual ~CoordinateMap_();
5   virtual double operator()
        (double y, double* yp = 0, double* ypp = 0)
    const = 0;
    virtual double Y(double x) const = 0;
};
```

A *coordinate vector* specifies how many points are to be used in a given spatial dimension; their bounds (between which they are evenly distributed) in the solver's coordinate space; and the mapping from solver space (points y_i) to model space (x_i).

```
────────────────── PDE.h ──────────────────
struct CoordinateVector_
{
    double yLow_, yHigh_;
    int n;    // n >= 2 if low != high
5   Handle_<CoordinateMap_> yToX_;
    map<Time_, double> rescalings_;

    void Save(Archive::Dst_& dst) const;
};
```

Note that `CoordinateVector_`, though it implements a `Save` method, is not `Storable_`; it is a concrete final `struct` which offers this method as a service to `Storable_` objects which may contain it.

One useful coordinate map is simply $x = \lambda \sinh(y/\lambda)$, which approaches a linear map (equivalent to the identity) as $\lambda \to \infty$. Clearly λ is the necessary member data for this mapping, but we construct it through the factory function

```
━━━━━━━━━━━━━━ PDEUtils.cpp ━━━━━━━━━━━━
  PDE::CoordinateMap_* PDE::NewSinhMap
     (double x_width,
      double dxdy_range)
  {
5    assert(IsPositive(x_width) && dxdy_range >= 1.0);
     double sinhMaxY = sqrt(Square(dxdy_range) - 1.0);
     return IsZero(Square(sinhMaxY))
            ? (CoordinateMap_*) new IdentityMap_
            : new SinhMap_(x_width / sinhMaxY);
10 }
```

Of course, like any factory function, this lets us keep the class `SinhMap_` completely within its source file.

7.9.3 Coefficient Calculators

The coefficient calculators must obviously compute a coefficient which might be x-dependent; they must also state their x-dependence so that the solver can avoid wasteful recomputation of unchanging coefficients. For a matrix coefficient, we must state the x-dependence of each element of the matrix. Thus we write (still in `namespace PDE`)

```
━━━━━━━━━━━━━━━ PDE.h ━━━━━━━━━━━━━━
  static const size_t MAX_DIMENSIONS = 3;
  class Coeff_
  {
  public:
5    virtual ~Coeff_();
     typedef bitset<MAX_DIMENSIONS> x_dep_t;
  };

  class MatrixCoeff_ : public Coeff_
10 {
  public:
     virtual void Value
        (const Vector_<>& x,
         SquareMatrix_<>* value)
15   const = 0;

     virtual Matrix_<x_dep_t> XDependence() const = 0;
  };
  MatrixCoeff_* NewConstCoeff(const Matrix_<>& val);
```

```
20  class VectorCoeff_ : public Coeff_
    {
    public:
       virtual void Value
25        (const Vector_<>& x,
           Vector_<>* value)
       const = 0;

       virtual Vector_<x_dep_t> XDependence() const = 0;
30  };
    VectorCoeff_* NewConstCoeff(const Vector_<>& val);

    class ScalarCoeff_ : public Coeff_
    {
35  public:
       virtual void Value
          (const Vector_<>& x,
           double* value)
       const = 0;
40
       virtual x_dep_t XDependence() const = 0;
    };
    ScalarCoeff_* NewConstCoeff(double val);
```

The **Value** functions take an output parameter to be written on, rather than returning a result, which violates our normal preference for functional code; the repeated construction of vectors and matrices would be too slow. For the scalar, while there is no efficiency penalty, it seems worthwhile to keep the function signatures consistent.

These calculators are produced by the model, and know nothing of coordinate transformations. Thus they work completely in x-space. We also make available to all models the utilities

```
─────────────────── PDE.h ───────────────────
namespace PDE
{
    MatrixCoeff_* NewConstCoeff(const Matrix_<>& val);
    VectorCoeff_* NewConstCoeff(const Vector_<>& val);
5   ScalarCoeff_* NewConstCoeff(double val);
}
```

7.9.4 *Forward Induction*

The persistence of fully-explicit, usually trinomial, "trees" for solution of PDEs is a mixture of tragedy and farce. Even the most naive extension of Crank-Nicholson – with cross terms computed using a fully explicit method – is stable and substantially more accurate than even a carefully built tree. Trees may be useful in pedagogy, but have no place in the working world of pricing.

A common excuse given by practitioners of this sorry art is that implicit or semi-implicit methods, despite their other desirable properties, do not support forward induction. Forward induction is doubtless useful: it is needed to fit the discount curve in some models such as Black-Karasinski, and is used for pricing many options in a single PDE sweep in Dupire-style local vol models. But forward induction is far less difficult than is often thought, even for sophisticated PDE solvers.

Suppose we have a set of possible states x_i, and corresponding values $U_i^{(+)}$ at time t_+.[15] The task of a backward solver is to compute $U_i^{(-)}$ at some earlier time t_-; all the solvers we consider are linear in U, so we have $U_- = MU_+$ for some matrix M independent of U.

Next, let $G_i^{(\pm)}$ represent the state prices at time $t_\pm$. We know the present value of the payout is $G^{(-)} \cdot U^{(-)} = G^{(+)} \cdot U^{(+)}$; and this is true *for any payout U*. Thus $G^{(-)} \cdot MU^{(+)} = G^{(+)} \cdot U^{(+)}$ for any $U^{(+)}$, which requires $G^{(+)} = M^T G^{(-)}$.

For models with space-dependent coefficients, the forward Kolmogorov equation is complicated by many extra terms, and if we discretize it we cannot take advantage of the above relation. Separately discretizing the forward equation is a mistake; in fact, even *deriving* the forward Kolmogorov equation is a mistake. We should discretize the backward equation *only*, and then take the "numerical dual" using the above relation. (This does require some judgement in the choice of a boundary condition which is sensible for both directions; however, bear in mind that the forward evolution only makes explicit the boundary's preexisting effects on backward-induction pricing.) The result is that the forward step is as simple to implement as the backward – in fact, they share the vast majority of their code – and the resulting state prices are consistent with backward induction pricing to machine precision.

---

[15]Our notation is that of a single spatial dimension, but our analysis is not restricted to that case: the index i here simply enumerates all nodes in any number of dimensions.

7.10 American Monte Carlo

In order to simulate optimal exercise of a Bermudan option, we must make some statement about the future expected value along two available branches; usually one branch's value is a known exercise value, and we are concerned with the *continuation value* along the other branch.

In non-recombining or "bushy tree" methods, we essentially spawn a new Monte Carlo beginning at the node where we must make an exercise decision. This child simulation must spawn its own children when it reaches a later exercise date; thus the cost of simulation grows exponentially with the number of exercise dates, relegating this method to the world of thought experiments.

American Monte Carlo (AMC) methods seek to replace the computation of continuation values using child simulations with an estimate gleaned from the paths we have already run. Thus AMC is fundamentally about *estimation* of continuation values, and only incidentally about *optimization* of exercise decisions. As such, it relies on the independent variables of the estimator, which must be easily measured at each node of the simulation; these are called *observables*. In practice we prefer to estimate the *exercise gain*, which is simply the difference between the exercise and continuation values.

7.10.1 *Recursive Partitioning*

The well-known Longstaff-Schwartz AMC algorithm uses a very simple estimator, relying on a rich set of observables to achieve acceptable results. We can reduce the observables, thus simplifying the setup and computation at each node, with a more sophisticated estimation procedure.

One way to improve the estimator is to increase its locality through recursive partitioning of the path set, or "bundling." Here we sort according to the last observable, partition the path set into bundles, then use the next-to-last observable to sort and partition into smaller bundles, and so on. The algorithm proceeds back-to-front in order to make the resulting bundles more local in the observables nearest the front, effectively making those observables more important.

The partitioning code itself creates two outputs, a `key` by which the paths are ordered and a list of `breakpoints` within the key; thus `key[0]`, `key[1]`, ... `key[b-1]` are the indices of the paths in the first bundle, where `b` is the first breakpoint.

```
                            ─── AMC.cpp ───
    void Partition
        (const Matrix_<>& observables,
        const Vector_<int>& n_bundles,
        bool bundle_first,
5       Vector_<int>* key,
        list<int>* breaks)
    {
        const int nPaths = observables.Columns();
        *key = IndicesTo(nPaths);    // [0, nPaths)
10
        breaks->clear();
        breaks->push_back(0);
        breaks->push_back(nPaths);
        const int minD = bundle_first ? 0 : 1;
15      for (int d = observables.Rows() - 1; d >= minD; --d)
        {
            int from = breaks->front();
            for (auto p = Next(breaks->begin());
                    p != breaks->end(); from = *p++)
20          {
                sort(&(*key)[from], &(*key)[*p],
                    ObservableLess_(observables, d));
                const int size = *p - from;
                const int nb = Min(size, n_bundles[d]);
25              for (int j = 1; j < nb; ++j)
                    breaks->insert(p, from + (size * j) / nb);
            }
        }
    }
```

This relics on the sorting predicate

```
                            ─── AMC.cpp ───
    struct ObservableLess  : binary_function<int, int, bool>
    {
        const Matrix_<double>& obs_;
        const int depth_;
5       ObservableLess_(const Matrix_<double>& obs, int d);
        bool operator()(int lhs, int rhs) const
        {
            return obs_(depth_, lhs) < obs_(depth_, rhs);
        }
10  };
```

7.10.2 *Biases*

The flag `bundle_first` above is useful because, once we have sorted and partitioned using all the other observables, we can use other methods for the one-dimensional problem within a single bundle. In particular, a non-parametric fit based on linear smoothing splines can capture more of the signal than a rigid estimator across bundles, while simultaneously enlarging the set of paths influencing the estimator; that is, it can reduce both *granularity bias* from the nonzero spatial extent of the basis of estimation and *small-sample bias* from the finite pathset size.

There is a third, more vicious bias with which we must also contend; the *lookback bias* resulting from the influence of a path's own future on its exercise decision. If the bundles are too small or the smoothing spline too flexible, the path's observed future in this particular pathset will dominate our estimate of the expected continuation value, and the prices we generate will be those of lookback options.[16] This leads us to consider the *jackknifed estimator* in which a path's own future is excluded from the estimation process for that path; however, this is not a panacea, since it does nothing to control the other two (downward) biases.

Consider a simplified model of the estimation procedure. For a given path, let μ be the true expected gain from exercise, and assume that our measurement of the gain (*i.e.*, the observed gain from the future of that path) is normally distributed. Thus our measured gain on a given path is $\mu + \epsilon_i$ where the ϵ's are i.i.d. $N(0,1)$ deviates. Without loss of generality we can assume the measurement variance is 1. We also assume the pathset is large so that we can ignore edge effects; this lets us simplify our notation by considering "path zero" within a pathset extending to both positive and negative indices.

The value realized on the path, including the option to exercise, can be written in terms of this error term as $(\mu + \epsilon_0)\mathbf{1}_{\hat{\mu}_0 > 0}$ where $\hat{\mu}_0$ is the estimated exercise gain on this particular path. If our estimator is linear, then in our model we can write it as $\mu + \beta\epsilon_0 + \gamma\epsilon_\perp$, where $\epsilon_\perp$ is uncorrelated with ϵ_0 and encapsulates the total effect of all other paths. This expectation can be computed with a simple rotation of variables, obtaining

$$\mu N\left(\frac{\mu}{\sigma}\right) + \frac{\beta}{\sigma}\phi\left(\frac{\mu}{\sigma}\right)$$

where ϕ is the normal density function and $\sigma^2 \equiv \beta^2 + \gamma^2$. The true value is

---

[16]Worse, in fact, since the payout need not be discounted over a lookback period.

of course μ^+; the former term is always less than this due to small-sample bias, while the latter positive term is the lookback bias.

Since we are interested in the total bias over the whole range of paths, we next integrate over μ. The result is $\beta - \frac{1}{2}\sigma^2$ – to which *the total bias introduced into pricing* is proportional. This shows the possibility of balancing the lookback bias against the small-sample bias, especially in the region where k is significantly below 1, where both increase in importance.

If we further assume that there is only one observable, and that the paths are equidistant in this independent variable, then our linear smoothing spline reduces to an average weighted by a double-exponential with decay constant k:

$$\hat{y}_0 = \mu + \frac{1-k}{1+k}\sum_{i\in\mathcal{Z}}\epsilon_i k^{-|i|} = \mu + \frac{1-k}{1+k}\epsilon_0 + \frac{k}{1+k}\sqrt{\frac{2(1-k)}{1+k}}\epsilon_\perp.$$

For brevity we define $H \equiv (1-k)/(1+k)$, so in our notation

$$\beta = H, \qquad \gamma = (1-H)\sqrt{\frac{H}{2}};$$

similarly, the jackknifed estimator has

$$\beta = 0, \qquad \gamma = \sqrt{\frac{H}{2}}.$$

Now suppose that we use a linear combination of the two estimators, weighting the first by w_L (the *lookback weight*). The combined estimator has

$$\beta = w_L H, \qquad \gamma = (1-w_L H)\sqrt{\frac{H}{2}};$$

the bias-cancelling condition becomes

$$w_L = \frac{1 - \sqrt{\frac{2}{2+H}}}{H},$$

which rapidly approaches $w_L = \frac{1}{4}$ as $H \to 0$ (the limit of rigid smoothing splines). While this result has been derived under highly restrictive assumptions, in practice it is applicable to real problems and greatly increases the power of estimation-based deciders.

This page intentionally left blank

Chapter 8

Schedules

When we attempt to apply mathematical techniques to financial instruments, there is a constant tension between the "pure" mathematical regime – of continuously compounded rates, uniform daycounts, and so forth – and the idiosyncratic real world, with its profusion of entrenched conventions. This is particularly pronounced in fixed income, but is an issue for any underlying. The unifying goal of schedule code is to bridge this divide by providing to the mathematical side the tools to encapsulate these conventions and quarantine them so we can get on with our real work. We make no pretense that schedules are interesting, but they are crucial.

8.1 Enumerated Switches

The details of a schedule based contract, such as a Libor swap, are determined by a combination of choices: roll convention, daycount, and so forth. For example, the fixed coupon rate is converted to a cash payment amount for one period by multiplying by the notional amount and by a daycount fraction, which is (probably) computed by applying some accepted "daycount basis" to the accrual period start and end dates. Thus our parametrization of a swap must indicate the method for this computation.

This calls for a machine-generated `ENUM` type as in Sec. 3.8 – a class constructible from a string. But there is a complication: we must be prepared for new methods to be added. For instance, when we expand our emerging-markets operations we must begin dealing with Brazilian swaps whose daycount is 1/252 of the number of Rio de Janeiro business days in the interval. We describe such an *extensible enumeration* in another mark-up file:

```
────────────────── DayBasis.enum.if ──────────────────
  ' Coupon daycount basis.
  ALTERNATIVE NOT_SET
  ' A placeholder; cannot be used for anything.
  ALTERNATIVE ACT_365F ACT/365F ACT_365FIXED ACT/365FIXED
5 ' Always uses a 365-day year
  ALTERNATIVE ACT_365L ACT/365L ACT_365LONG ACT_365LONG
  ' Sometimes uses a 366-day year
  ALTERNATIVE ACT_360 ACT/360 MONEY ACTUAL/360
  ALTERNATIVE ACT_ACT ACT/ACT ACTUAL/ACTUAL
10 ALTERNATIVE BOND 30_360 30/360
  MEMBER typedef double result_type;
  MEMBER double operator()(long start_date, long end_date)\
      const;   // inadequate; see below
```

Because we have not declared this type to be **SWITCHABLE**,[1] the generated code will look somewhat different from that of the enumerations shown in Sec. 3.8.

```
────────────────── DayBasis.cpp ──────────────────
  class DayBasis_
  {
  public:
      class Extension_
5     {
      public:
          virtual ~Extension_();
          // Implement DayBasis_ interface
          virtual const char* String() const = 0;
10        virtual double operator()(long start_date,
              long end_date) const = 0;
      };
  private:
      enum EDayBasis
15    {
          NOT_SET,
          ACT_365F,
          ACT_365L,
          ACT_360,
20        ACT_ACT,
          BOND,
          _EXTENSION,
          _N_VALUES
      } val_;
```

[1]We can define a **CLOSED** tag, to label enumerations which are neither switchable nor extensible, but this is seldom useful.

```
25    Handle_<Extension_> other_;
      const Extension_& Extension() const;
      DayBasis_(EDayBasis val)
            : val_(val) {assert(val < _EXTENSION);}
      DayBasis_(const Handle_<Extension_>& imp)
30            : val_(_EXTENSION), other_(imp) {}
      friend bool operator==(const DayBasis_& lhs,
         const DayBasis_& rhs);
      friend vector<DayBasis_> DayBasis_ListAll();
   public:
35    explicit DayBasis_(const String_& src);
      const char* String() const;
   // Idiosyncratic (hand-written) members:
      double operator()(long start_date, long end_date)
         const;
40 };
```

Here the interface generator has to be a bit sneaky to generate the
`Extension_` base class: it notices that one of the members ends with `const`;
and uses that information to generate a virtual member of `Extension_`. We
need this extra step to avoid machine-generating invalid code like

```
virtual typedef double result_type - 0;
```

If we are determined to avoid this hack, we can require the mark-up to state
explicitly which members are not to be transplanted to the `Extension_`
(e.g., by using `MEMBER:FINAL` to mark them).

8.1.1 *Groundwork for Extensibility*

We must implement a constructor from string for `DayBasis_`, but we cannot
yet know what other values may be declared elsewhere. Also we have rashly
promised to `ListAll` the values.

Thus we must supply a run-time registry for recognition functions and
the corresponding canonical names. We use Meyers singletons:

DayBasis.cpp

```
  namespace
  {
     vector<DayBasis_>& DayBasis_List()
     {
5       static vector<DayBasis_> RETVAL;
        return RETVAL;
     }
```

```
     map<String_, Handle_<DayBasis_::Extension_> >&
10      DayBasis_Extensions()
     {
        static map<String_,
           Handle_<DayBasis_::Extension_> > RETVAL;
        return RETVAL;
15   }

     void DayBasis_MustFail(const String_& test)
     {
        try
20      {
           DayBasis_ temp(test);
        }
        catch (Exception_&)
        {
25         return;
        }
        throw Exception_("'" + test + "' is already a "
           "DayBasis_, can't be declared again");
     }
30 }   // leave local namespace
```

This is the registry to which extension classes will add themselves. Note that this implementation assumes a limited set of exact (modulo case, whitespace and underscores) matching strings for each enumerated value, rather than a more generalized notion of patterns. This is by design, as the risk of false positives makes uninhibited pattern-matching too dangerous for enumerated switches in production.

The awkward code of **MustFail** arises from our need to throw an exception if and only if the construction of **temp** succeeds; this is the surest test of a name collision.

DayBasis.cpp

```
void DayBasis_RegisterExtension
   (const vector<String_>& names,
   const Handle_<DayBasis_::Extension_>& val)
{
5  Require(0, DayBasis_List().empty(), "Can't register"
      "new DayBasis after enumerating all values");
   assert(!val.Empty());
   DayBasis_MustFail(val->String());
   for (auto pn = names.begin(); pn != names.end(); ++pn)
10    {
```

```
        // check that this name is not already recognized
        DayBasis_MustFail(*pn);
        const String_ key = String::Condensed(*pn);
        DayBasis_Extensions()[key] = val;
15  }
    // check that two-way string conversion works
    DayBasis_ check(val->String());
}
```

The final check ensures that the **String** method implemented by **val** does indeed produce a valid **DayBasis_**; the initial call to **MustFail** assures us that it is our own **val** that is being used.

External users may wish to list all possible day bases at runtime – *e.g.*, to populate a drop-down list in a user interface. Our support for this is based on the function

```
———————————— DayBasis.cpp ————————————
vector<DayBasis_> DayBasis_ListAll()
{
    vector<DayBasis_> retval;
    set<String_> exists;
5   for (auto ii = DayBasis_::EDayBasis_(0);
            ii != DayBasis_::_EXTENSION; ++ii)
    {
        retval.push_back(DayBasis_(ii));
        exists.insert(DayBasis_(ii).String());
10  }
    const map<String_, Handle_<DayBasis_::Extension_> >&
            more = DayBasis_Extensions();
    for (auto pe = more.begin(); pe != more.end(); ++pe)
    {
15      if (!exists.count(pe->second->Name()))
        {
            retval.push_back(DayBasis_(pe->second));
            exists.insert(pe->second->Name());
        }
20  }
    return retval;
}
```

It is clear that all the above code can be machine-generated: in fact, it is generated by mechanical substitution of the string **"DayBasis"** into a prefabricated block of pseudocode.

8.1.2 *30E/360 ISDA, ACT/ACT ISMA*

Some conventions do not fit naturally into this framework. 30E/360 ISDA, used by some German bonds, treats the last period differently from others: a period ending on the last day in February will be treated as February 30 for daycount purposes, *but* only if it is not the last coupon period. In actual/actual ISMA, used by some U. S. Treasury bonds, the period daycount cannot be determined without reference to the nominal start and end dates and the coupon frequency.

If we are to support these conventions, we have two unpalatable choices: we can replace the simple operator() above with

DayBasis.enum.if

```
MEMBER double operator()(long start_date, long end_date,\
    bool is_last, long nominal_start, long nominal_mat, \
    int cpn_months) const;
```

and then ensure that all daycounts have this information whenever they might need it. Alternatively, we might treat legs using these daycount conventions as nonstandard, and supply separate special-purpose constructors for them. We reluctantly adopt the first approach.

We sweeten the bitter pill somewhat by packaging the otiose information in a single argument,

DayBasis.h

```
  struct DayBasisContext_
  {
      bool isLast_;
      long nominalStart_;
5     long nominalEnd_;
      int couponMonths_;
  };
```

Now we can write the less voluble

DayBasis.enum.if

```
MEMBER double operator()(long start_date, long end_date,\
        const DayBasisContext_* context) const;
```

8.1.3 *BUS/252*

A similar problem is introduced by the presence of daycounts depending on a number of business days, which in turn depends on the holiday calendar used. This can be implemented in two ways: by making the daycount depend on an input holiday calendar (adding yet another argument to `operator()` above), or by observing that different holiday calendars create different day bases. We prefer the latter approach, which separates the day basis from the other uses of the holiday calendar in constructing a swap schedule. This obliges us to create a new day basis for each holiday schedule used with a Business/252 basis; however, such daycounting is widespread only in Brazil.[2]

We create a file describing this additional day basis:

```
─────────────────── Bus252.DayBasis.enum.if ───────────────────
ALTERNATIVE BRL_BUS_252 Brazil_BUS_252
'Brazilian daycount convention.
MEMBER double operator()(long start_date, long end_date,\
        const DayBasisContext_* context) const;
```

The `MEMBER` must be repeated because the code generator will not have access to `DayBasis.enum.if`. Now we have supplied enough information to generate code defining the extension class:

```
─────────────────────── Bus252.h ───────────────────────
   struct Bus252_DayBasis_ : public DayBasis_::Extension_
   {
       enum EBus252_DayBasis
       {
5         BRL_BUS_252,
          Bus252_N_VALUES
       } val_;
       Bus252_DayBasis_(EBus252_DayBasis val) : val_(val) {}
       const char* String() const;
10     double operator()(long start_date, long end_date,
           const DayBasisContext_* context) const;
   };
```

The registration is accomplished by our usual method, declaring a file-scoped static object whose constructor calls the registration function. Because Meyers singletons are initialized on the first call to the function containing their definition, we are not at risk of initialization-order bugs: the central registry will be created when it is needed.

```
─────────────────────── Bus252.cpp ───────────────────────
   struct _Bus252_DayBasis_Register_
   {
       _Bus252_DayBasis_Register_()
       {
5         Handle_<DayBasis_::Extension_> a_BRL_BUS_252
              (new Bus252_DayBasis_
```

---

[2]Where it is a legacy of hyperinflation.

```
                      (Bus252_DayBasis_::BRL_BUS_252));
           vector<String_> names_BRL_BUS_252;
           names_BRL_BUS_252.push_back("BRL_BUS_252");
10         names_BRL_BUS_252.push_back("Brazil_BUS_252");
           DayBasis_RegisterExtension
                 (names_BRL_BUS_252, a_BRL_BUS_252);
       }
   };
15 static const _Bus252_DayBasis_Register_
           The__Bus252_DayBasis_Register;
```

String is easy to machine-generate as well. The only task that remains for us it to implement the relevant **operator()**; this functionality will then be available to any user, once the code is loaded.

The code to work with an enumerated type does not change when that type is made extensible; this is a crucial advantage, letting us open up our system in ways we may not originally have anticipated. We should as a rule avoid making enumerations **SWITCHABLE**, but that is simply the good coding practice of minimizing unnecessary assumptions.

8.1.4 *Other Enumerations*

This will be our preferred method for defining any list of user choices. The uniform interfaces of the machine-generated code support the reliable generation of code in other areas – *e.g.*, at the public interface, where we will take **ENUM** inputs. We will also enjoy the fringe benefits of code generation, particularly simultaneous generation of user help pages.

Though we have focused here on **DayBasis_**, other enumerated switches follow precisely the same pattern. Our job is to distill the use of the enumeration into a few C++ function signatures, which are declared as **MEMBER**s in mark-up descriptions and implemented by hand. Common enumerated types for schedule generation include

- Business-day convention (applied if a nominal period end is not a business day). This presents one subtlety, like that in the discussion of 30E/360 day basis above; sometimes accrual dates are not adjusted, though payment dates always are. Is this a feature of the leg, or of the roll convention?
- Roll type (for swaps whose period end dates are not computed from the start date, such as IMM swaps).
- Frequency, or period length. Because these are two forms for the same

information, conversion to and from integers must be very explicit (does 1 mean monthly or annual)?

- Averaging frequency, for averaging basis swaps.

8.2 Holidays

8.2.1 *Cities*

Each jurisdiction in which a trade might be consummated will define a *holiday calendar*, indicating which days are not *good business days* there. A jurisdiction, in this role, is called a *holiday center* or simply a *city*.

The set of holidays for a city is subject to change, sometimes at fairly short notice, as new special occasions are deemed worth celebrating. Also, an expanding international business will routinely find itself involved in new jurisdictions, necessitating the addition of new holiday calendars. Maintenance of these holiday calendars is not a natural or rewarding job for front-office quants.

Thus holidays are best regarded as configuration data, which we will read at library load time. The data are fairly simple: for each of some (fairly large) set of city codes, we will supply a set of all holidays over some (fairly wide) date range, from `Date::MINIMUM` through `Date::MAXIMUM`. Holidays may be more compactly represented by rules (e.g., 25 December is always a London holiday when on a weekday), but this rule-based representation is best kept encapsulated within the configuration loader, or within a completely separate application which generates data to be loaded. For our purposes, we will expect the library to obtain a `map<String_, vector<long> >`.

For efficiency, we will store this data in two parts: a map of cities to integer indices, and an index vector of holidays.

———————————— *HolidayData.h* ————————————

```
struct HolidayData_
{
    vector<String_> centers_;
    map<String_, int> centerIndex_;
    vector<vector<long> > holidays_;
    bool IsValid() const;
    void Swap(HolidayData_* other);
};
HolidayData_& TheHolidayData()
{
    static HolidayData_ RETVAL;
```

```
      return RETVAL;
}
```

This representation (in a local namespace) lets us convert back and forth between cities and their indices. The `IsValid` code checks class invariants, such as correspondence between the keys of the `centerIndex_` and the `centers_`.

———— *HolidayData.cpp* ————
```
void AddHolidayCenter(const String_& city,
    const vector<long>& holidays)
{
    assert(TheHolidayData().IsValid());
 5  HolidayData_ temp(TheHolidayData());
    REQUIRE0(!temp.centerIndex_.count(city),
        "Duplicate holiday center '" + city + "'");
    temp.centerIndex_[city] = temp.centers_.size();
    temp.centers_.push_back(city);
10  assert(ContainsNoWeekends(holidays));
    assert(IsMonotonic(holidays));
    temp.holidays_.push_back(holidays);
    TheHolidayData().Swap(&temp);
    assert(TheHolidayData().IsValid());
15 }
```

Use of the copy-swap idiom protects us from the unlikely event of an exception during the insertion process, which would make the library unusable.

8.2.2 *Holiday Sets*

It is quite common for dates to depend on multiple holiday schedules. For instance, the Libor start date for a given fixing date uses both New York and London holidays; and an offshore corporate counterparty to a swap is likely to ask that swap payments be rolled to good business days in their city of operations.

Thus the fundamental object used in schedule generation is a *holiday set*, which can in special circumstances be empty, of holiday centers. This is constructed from a list of holiday center codes; we prefer a space-separated list (which can then be included in a larger comma-separated list on occasion), but our parser will accept commas as well, and possibly other separators. A blank string should not be interpreted as "no holidays", which is really a rare usage and invalid in many situations; instead, some special string like `"NONE"` should be required. For the same reason, we do not provide the holiday set with a default constructor.

Our class for holiday sets will store the indices of the centers in `TheHolidayData`; this is sufficient for a minimal implementation.

```
──────────────── Holiday.h ────────────────
class Holidays_          // minimal
{
    vector<int> centers_;
    friend class CountBusDays_;
public:
    Holidays_(const String_& src);
    String_ String() const;
    bool IsHoliday(long date) const;
};
```

The implementation of `IsHoliday` proceeds as one would expect:

```
──────────────── Holiday.cpp ────────────────
bool Holidays_::IsHoliday(long date) const
{
    for (auto pc = centers_.begin();
        pc != centers_.end(); ++pc)
    {
        if (BinarySearch(TheHolidayData().holidays_[*pc],
            date))
            return true;
    }
    return false;
}
```

```
──────────────── Holiday.cpp ────────────────
String_ Holidays_::String() const
{
    if (centers_.empty())
        return String_();
    auto toName = AsFunctor(TheHolidayData().centers_);
    return String::Unsplit
            (Apply(toName, Unique(centers_)), ' ');
}
```

By implementing `Holidays_` in the same source file as `AddHolidayCenter`, we ensure that `TheHolidayData` is accessible from both while remaining insulated from non-holiday code.

More complex tasks, like a count of business days in an interval, can in principle be implemented as nonmember functions calling `IsHoliday` repeatedly. For performance, we may wish to support this more directly. The

ability to count holidays in an interval could be supported by pre-merging
the holiday lists in a new member `vector<long> Holidays_::vals_`:

```
// deprecated version
int Holidays_::CountHolidays(int begin, int end) const
{
    auto pf = LowerBound(vals_, begin);
    return lower_bound(pf, vals_.end(), end) - pf;
}
```

This counts the elements of `vals_` lying in the half-open interval [begin,
end), echoing the standard use of those names. Note the use of both the
standard and the container-level versions of lower bound, as explained in
Sec. 4.2.

But populating `vals_` on construction is likely a pessimization; it requires at least a vector copy for each holiday center in the set. We
could make `vals_` mutable, and ensure it is initialized at the start of
`CountHolidays`: this is reasonably efficient but reflects the poor practice
of mixing states (*i.e.*, initialized and uninitialized) within the same type;
see Sec. 2.3.

It is better to count business days through a separate class:

─────────────────────────── *Holiday.h* ───────────────────────────
```
class CountBusDays_
{
    Handle_<vector<long> > hols_;
public:
    CountBusDays_(const Holidays_& src);
    int operator()(long begin, long end) const;
};
```

Now, instead of

```
Holidays::CountBusDays(holidays, begin, end)
```

we will write

```
CountBusDays_(holidays)(begin, end)
```

which is not substantially more complex.

The class stores the vector `hols_` in a handle, rather than by value,
because its constructor will maintain a cache of known combinations:

```
————————————— Holiday.cpp —————————————
CountBusDays_::CountBusDays_(const Holidays_& src)
{
    static map<vector<int>, Handle_<vector<long> > >
        TheHolidays;
5   Handle_<vector<long> >& hols
            = TheHolidays[Unique(src.centers_)];
    if (hols.Empty())
    {
        vector<long> h;
10      for (auto pc = src.centers_.begin();
                    pc != src.centers_.end(); ++pc)
            Append(&h, TheHolidayData().holidays_[*pc]);
        hols.reset(new vector<long>(Unique(h)));
    }
15  hols_ = hols;
}
```

Here we have assumed access to the internals of `Holidays_`; thus `CountBusDays_` should be a **friend** of that class.[3]

This is the main time-intensive application of holiday schedules. We could in principle optimize other functions (like finding the next good business day) in the same way, but there is no real gain from this. It is better to keep the classes simple and supply nonmember functions as needed, though we might add a member `CountBusDays_::IsHoliday` to that `CountBusDays_` can be used as an optimized holiday set.

8.3 Currencies

Besides being the means of payment, currencies also supply default conventions for many trades: once I know that a Libor swap is denominated in USD, I can surmise that its fixed side is semiannual with 30/360 day basis and New York payment holidays, and so on. Of course, these may be overridden for a particular trade, but it is necessary to know the defaults: they complete the definition of market observable indices and of quoted yield curve build instruments, since the markets for nonstandard trades are much less deep, less liquid and less public.

For every currency, we must be able to look up a wide variety of facts. Since these facts are of different types, from **int** on up, storing and indexing

---

[3]Friend status is not completely necessary; we could index `TheHolidays` by a string key, obtained from `src.String`, with only a minor efficiency penalty.

them is a surprisingly deep problem. We must decide on

- Order of indexing – is the currency the accessor's first or last argument?
- Extensibility – can we store and fetch conventions whose existence was not foreseen?
- Typing – where does the conversion from text data occur?
- Syntax – can we fetch a single fact without extra type clutter?

The syntax and typing constraints turns out to be surprisingly strong. For example, we must type some permutation of Ccy::Conventions, "USD", SwapPayHolidays and get back a Holidays_ without any further ado. Thus the type information must come from the name of the thing requested: so it must identify a member datum, function, or concrete class.

The desire for extensibility dictates that the currency should be the last, not the first, argument; then new quantities can be introduced by introducing new functions in a namespace, which can be done anywhere. This also improves physical code structure, since different conventions need not all be tied together.

Thus we end up writing

```
Ccy::Conventions::SwapPayHolidays()("USD");
```

which is not so different from our first guess.

The return value of SwapPayHolidays should expose a writing interface as well (otherwise we must create a separate writing interface elsewhere). This leads us to the following implementation, which we place in **namespace Ccy**:

```
─────────────── CurrencyData.h ───────────────
   template<class T_> class OneFact_ : noncopyable
   {
   public:
       virtual ~OneFact_();
5      virtual const T_& operator()
            (const String_& ccy) const = 0;

       class Writer_ : noncopyable
       {
10     public:
           virtual void SetDefault(const T_& val);
           virtual void operator()
               (const String_& ccy, const T_& val);
       };
```

```
15    virtual Writer_& XWrite() const = 0;
};
```

Thus we store facts with code like

```
Ccy::Conventions::SwapPayDelay().XWrite().SetDefault(2);
Ccy::Conventions::SwapPayDelay().XWrite()("GBP", 0);
```

8.3.1 *Internals*

The above interface will clearly be supported by a singleton data holder, a dictionary plus default value, whose definition can be localized to a single source file.

```
――――――― CurrencyData.h ―――――――
template<class T_> struct CcyDependent_
{
    Handle_<T_> background_;
    map<String_, Handle_<T_> > specific_;
};
```

We hold the **specific_** values in **Handle_**s so that our code will work for classes – like **DayBasis_** – which lack a default constructor.

Our implementation of a currency-dependent fact will store a **CcyDependent_**. We implement the writer by providing access directly to this data:

```
――――――― CurrencyData.cpp ―――――――
template<class T_> class XFactWriterImp_
      : public OneFact_<T_>::Writer_
{
    CcyDependent_<T_>& data_;    // we do not own
public:
    XFactWriterImp_(CcyDependent_<T_>& dat) : data_(dat) {}

    void SetDefault(const T_& val)
    {   data_.background_.reset(new T_(val));    }
    void operator()(const String_& ccy, const T_& val)
    {   data_.specific_[ccy].reset(new T_(val));    }
};
```

We use a level of indirection to ensure that, when we construct our fact holder, the writer is constructed with a valid reference to the data.[4]

---

[4]We could take the writer off the heap and rely on order of initialization, but the gains are insignificant.

```
———————————————— CurrencyData.cpp ————————————————
template<class T_> class OneFactImp_ : public OneFact_<T_>
{
    CcyDependent_<T_> vals_;
    scoped_ptr<typename OneFact_<T_>::Writer_> writer_;
public:
    OneFactImp_()
    {   writer_.reset(new XFactWriterImp_<T_>(vals_));    }

    const T_& operator()(const String_& ccy) const
    {
        auto pc = vals_.specific_.find(ccy);
        if (pc != vals_.specific_.end())
            return *pc->second;
        REQUIRE0(!vals_.background_.Empty(),
                 "No default for '" + ccy + "'");
        return *vals_.background_;
    }
    Writer_& XWrite() const {return *writer_;}
};
```

Now each function in `Ccy::Conventions` returns a Meyers singleton of
this type. For each convention we need only one brief and simple function:

```
———————————————— CurrencyData.cpp ————————————————
const OneFact_<Holidays_>&
      Ccy::Conventions::SwapPayHolidays()
{
    static const OneFactImp_<Holidays_> RETVAL;
    return RETVAL;
}
```

This code will be called at library load time by a configuration reader, which
will read facts from some accepted source (*e.g.*, a database or configuration
file) and write them into the store for that session. None of these methods
are currency-specific; they will work equally well to store, *e.g.*, parameters
of commodity futures contracts.

8.4   Increments

A good part of the fixed income business revolves around incrementing
dates; by months, by calendar or business days, and so on. This is such a

general concept that it is worth embodying in its own code.

```
──────── DateIncrement.h ────────
namespace Date
{
    class Increment_ : noncopyable
    {
    public:
        virtual ~Increment_();
        virtual long FwdFrom(long date) const = 0;
        virtual long BackFrom(long date) const = 0;
    };

    long operator+(long date, const Increment_& inc)
        {return inc.FwdFrom(date);}
    long operator-(long date, const Increment_& inc)
        {return inc.BackFrom(date);}
}
```

We distinguish two types of increment:

(1) Increment by some number of intervals;
(2) Increment to the next (or previous) of some set of dates.

The latter category includes, e.g., advancing to the next quarterly IMM date. The best solution is an enumerated list describing which special dates allow this, which we maintain in another mark-up file:

```
──────── SpecialDays.enum.if ────────
'Date families supporting jump to-next
ALTERNATIVE IMM IMM3 IMM_QUARTERLY
'Quarterly IMM dates.
ALTERNATIVE IMM1 IMM_MONTHLY
'Monthly IMM dates.
ALTERNATIVE CDS CDS3 CDS_QUARTERLY
'Quarterly CDS standard maturities.
ALTERNATIVE CDS1 CDS_MONTHLY
'Monthly CDS standard maturities.
ALTERNATIVE EOM
'End of a month.
MEMBER long Next(long base) const;
MEMBER long Previous(long base) const;
```

Other sets of special days might be added, either in this file or in extensions (see Sec. 8.1). The implementation of one type of increment is now straightforward:

```
─────────────── DateIncrement.cpp ───────────────
class IncrementNextSpecial_ : public Increment_
{
    SpecialDays_ days_;
    long FwdFrom(long d) const {return days_.Next(d);}
5   long BackFrom(long d) const {return days_.Previous(d);}
public:
    IncrementNextSpecial_(const SpecialDays_& d);
};
```

The code of **SpecialDays_::Next** will also support Libor futures rates, allowing us to compute the IMM date in any given month by looking forward from the end of the previous month.

The other class of increment is also supported by an enumerated list, this time of step sizes:

```
─────────────── DateStepSize.enum.if ───────────────
'Date steps supporting multistep increment
ALTERNATIVE CALENDAR_DAY CD
'Advance by calendar days
ALTERNATIVE BUSINESS_DAY BD
5   'Advance by business days
ALTERNATIVE MONTH M
'Advance by months
ALTERNATIVE YEAR Y
'Advance by years
10  MEMBER long operator()(bool fwd, long from, int n_steps, \
        const Holidays_& hols) const;
```

We use a boolean flag to distinguish forward from backward steps, rather than a signed integer **n_steps**; this allows us to use **n_steps == 0** to mean forward or backward adjustment to a good business day (with no change if the **from** date is already a good business day). We might extend this enumeration with additional alternatives overriding the roll convention (e.g., to allow rolling into a new month, which is nonstandard).

An increment now specifies how many steps to take of a given type, and also gives the holiday calendar to use:

```
─────────────── Increment.cpp ───────────────
class IncrementMultistep_ : public Increment_
{
    int nSteps_;
    DateStepSize_ stepBy_;
5   Holidays_ hols_;
    long FwdFrom(long d) const
```

```
         {return stepBy_(true, d, nSteps_, hols_);}
      long BackFrom(long d) const
         {return stepBy_(false, d, nSteps_, hols_);}
10 public:
      IncrementMultistep_(int n, const DateStepSize_& d,
         const Holidays_& h);
 };
```

Now building a grammar for increment specification is very simple, because each increment either begins with a number (of steps to take), or is a `SpecialDays_`. We need to specify two features of the grammar; a separator between the `DateStepSize` and the (optional) `Holidays`, and a separator between multiple increments. The latter help to avoid an unnecessary distinction between single and compound increments; for example, "the next London business day after the next New York and London business day after (a day)" is a reasonable specification of an increment, and we should treat it as such. We prefer grouping brackets (`[]`) for the former separator, and `&` for the second; thus we would write such an increment as `1BD[NY LON]&1BD[LON]`. Note the left-to-right application; date increments are not commutative.[5] We support compound increments by defining a new type

DateIncrement.cpp
```
class IncrementCompound_ : public Date::Increment_
{
   Vector_<Handle_<Increment_> > vals_;
   ...
```

which lets user code effectively handle compound increments without always thinking about them.

8.5 Legs

A *swap* is the repeated scheduled exchange of one kind of payment for another; but these exchanges often are not synchronous or even one-for-one. For example, the standard USD Libor swap exchanges quarterly Libor for a semiannual fixed payment. Thus the individual leg, not the swap, is the building block on which we should concentrate.

We must distinguish the schedule of payments from the determination of the payment amounts. The former is described by an *accrual schedule*,

---

[5]This is the usual cause of "Thursday bugs."

often loosely called a *leg schedule*, which is in turn a sequence of *accrual periods*:

```
―――――――――――――― AccrualPeriod.h ――――――――――――――
  struct AccrualPeriod_
  {
      long startDate_;
      long endDate_;
5     DayBasis_ couponBasis_;
      double dcf_;    // redundant but handy
      Handle_<DayBasisContext_> context_;
      bool isStub_;    // used in forming Libor rates
  };
```

Now we must describe the *coupon rate*, which is multiplied by `dcf_` to give the amount paid. This is a surprisingly nontrivial problem. We might imagine a base class with a virtual `CouponRate` function; but this would place the burden of rate computation at the very low level of leg periods, not at the level of yield curves and models where it belongs. Also, we cannot predict in advance all the methods for fixing a coupon rate, nor the subsidiary data each will require; thus we need the flexibility of a full-fledged object. This leads us to the somewhat unsatisying

```
――――――――――――――― CouponRate.h ―――――――――――――――
  struct CouponRate_ : noncopyable
  {
      virtual ~CouponRate_() {}
  };
```

We will be obliged to construct free functions which discriminate different types of `CouponRate_` by `dynamic_casting`.

In particular, we will immediately define the common rate types

```
――――――――――――――― CouponRate.h ―――――――――――――――
  struct FixedRate_ : CouponRate_
  {
      double rate_;
      FixedRate_(double rate) : rate_(rate) {}
5 };

  struct LiborRate_ : CouponRate_
  {
      Time_ fixDate_;
10    String_ ccy_;
      String_ tenor_;
      LiborRate_(long fix_date, const String_& ccy,
```

```
        const String_& tenor);
};
```

We have chosen to make the `LiborRate_` contain its fixing date, rather than relying on obtaining it from the leg at the time of computation. This lets us make the rate computation fully independent.

Now we can build a leg by defining, for each period, the accrual terms and rate fixing terms:

```
────────── Period.h ──────────
struct LegPeriod_
{
    Handle_<AccrualPeriod_> accrual_;
    Handle_<CouponRate_> rate_;
    long payDate_;
};
```

This `struct` is designed to be held in vectors. The `accrual_` is held in a `Handle_` because `AccrualPeriod_` has no default constructor. We could provide a sorting relationship and use `set` instead of `Vector_`, but this has negative practical value. The periods of a single leg are generated in order and stay that way; and we will later see methods which combine multiple legs, without need for an ordering.

8.5.1 *Stubs*

For first- and last-period stub rates on Libor swaps, we must usually interpolate between published Libor fixings to generate the paid rate. If we write

```
struct InterpLiborRate_ : CouponRate_   // deprecated
{
    Handle_<LiborRate_> r1_, r2_;
    double w1_;    // and w2 = 1-w1
};
```

then we have a special-purpose structure which still must be supported wherever `LiborRate_` is.

Instead, we should write

```
────────── CouponRate.h ──────────
struct SummedRate_ : CouponRate_
{
```

```
    Vector_<pair<double, Handle_<CouponRate_> > > rates_;
};
```

This supports rates with margin and float-float spreads as well, with no substantial complexity burden. We use a **Vector_** to hold any number of rates for interpolation, rather than restrict to exactly two; the likely alternative is the case of only one rate (*e.g.*, a rate transformed by a non-unit gearing). Also see Sec. 11.5 for more on queries of **CouponRate_s**.

8.5.2 *Build from Parameters*

Leg schedules are not directly specified in the terms of a trade; they are worked out from higher-level parameters. These parameters, rather than the leg itself, should be **Storable_**.

There is some question whether to store parameters which agree with the defaults for the leg's currency; for example, a USD Libor leg is quarterly by default, so should we store the coupon period for a quarterly floating leg? If we do so, then we are protected in case the default changes, since existing trades have all their details stored. But in storing these details, we will bloat the trade description; and, should we make any mistake as to conventions, this mistake will be enshrined in any booked trades. Thus we will adopt the policy of storing as little data as possible.

It is convenient to form a **Storable_** containing the information needed to form a swap leg. We describe its object layout in mark-up[6]:

```
——————————— LegScheduleParams.1.storable.if ———————————
    BUILDS Handle_<LegScheduleParams_>
    DATE start_date
    ?DATE mat_date
    ?STRING tenor
5   ?STRING coupon_period
    ?STRING day_basis
    ?INTEGER roll_day
    ?STRING roll_direction
    ?STRING roll_special
10 '    E.g, for IMM rolls.
    ?BOOLEAN stub_at_end
    ?BOOLEAN long_coupon
    ?BOOLEAN pay_upfront
    ?STRING pay_holidays
```

---

[6]The 1 in the file name is a version number; if we change the data model, except by adding optional data, we must add a new version, while keeping the old version to read already-existing data.

```
15  *INTEGER pay_delay
    '    A single offset (number of business days) for all
    '    payments, or a vector of one entry per payment.
    CONDITION {mat_date.Known() != tenor.Known()} \
       {Must specify mat_date or tenor, not both}
```

This lets us machine-generate a **struct** (in **namespace Data**) containing the above members:

```
——————— LegScheduleParams.h ———————
struct LegScheduleParams_v1_
{
    long start_date;
    Maybe_<long> mat_date;
5   Maybe_<String_> tenor;
    Maybe_<String_> coupon_period;
    Maybe_<String_> day_basis;
    Maybe_<int> roll_day;
    Maybe_<String_> roll_direction;
10  Maybe_<String_> roll_special;
    Maybe_<bool> stub_at_end;
    Maybe_<bool> long_coupon;
    Maybe_<bool> pay_upfront;
    Maybe_<String_> pay_holidays;
15  Vector_<int> pay_delay;

    void Save(Archive::Dst_& dst) const;
    Handle_<LegScheduleParams_> Build
           (Archive::Built_* built) const;
20 };
```

Our **class LegScheduleParams_** will contain the same data, though our code generation scheme does not permit us to share the class definition.

Now we can create a generator for a fixed leg. We will collect such functions in namespace LegBuild.

```
——————— LegSchedule.h ———————
Vector_<LegPeriod_> Fixed
    (_ENV, const String_& ccy,
     const LegScheduleParams_& schedule,
     const RecPay_& rec_pay,
5    const Vector_<>& notional,
     const Vector_<>& coupon,
     const NotionalExchange_& exchange);
```

The **RecPay_** and **NotionalExchange_** classes are machine-generated enumerations, as in Sec. 8.1. The `notional` and `coupon` may be single numbers, or vectors with one entry per coupon period. For the latter, we must be able to report an error when the input vector has the wrong size; thus we take the `_ENV` input to support error messaging. The parameters of the schedule are taken from the input `schedule` when possible; if they are not provided there, then defaults are fetched as described in Sec. 8.3.

Since this is sufficient information to build a fixed leg, it forms our definition of a fixed leg trade; see Sec. 11.4.

We can similarly build a Libor leg:

```
─────────────────────────── LegSchedule.h ───────────
  Vector_<LegPeriod_> Libor
     (_ENV, const String_& ccy,
      const LegScheduleParams_& schedule,
      const RecPay_& rec_pay,
5     const Vector_<>& notional,
      const Vector_<>& margin,    // empty for none
      const Holidays_* fixing_holidays,
      const Vector_<int>& fixing_delay,
      const NotionalExchange_& exchange);
10
```

The only subtle feature of this code is the default behavior for omitted arguments. An empty input `margin` means no margin, as does the vector [0]. But an empty input `fixing_delay` means that the fixing delay should be the default for the given `ccy`; a zero fixing delay must be explicitly specified. Omitting `fixing_holidays` likewise means that the currency default should be used.

8.5.3 *CDS*

Credit default swaps have their own conventions, separate from those of Libor swaps; in particular, their maturities are always rolled to common (usually quarterly) nominal end dates to facilitate netting.

The premium leg of a CDS (payments made by the buyer of protection) is well described as a sequence of `LegPeriod_`s, but in pricing we must include the possibility of a partial and early coupon payment due to default within a period.

For a standard CDS, the division of the protection leg into periods has no financial consequence; we need only to specify the amortization schedule (of which the protection end date is a special case). Thus a CDS trade at the time of pricing can consist of a leg of premium payments, plus a `PWC_` giving the protection amount as a function of time.

8.5.4 *Inflation Instruments*

Inflation instruments introduce a different complication: the "inflation rate" on which payments are based is computed from a ratio of two price-index fixings. These indices are generally published only at one-month intervals, so an inflation leg must specify the fixing to be taken.

As a rule, we take the period start and end dates, subtract an "observation lag," and then choose whichever fixings fall in the same month as the resulting adjusted dates. This describes a "year-on-year" inflation rate; a less usual variant uses the "zero-coupon" inflation rate, where the first fixing (the denominator of the fixings ratio) is always tied to the leg's start date.

The particular date of the month on which inflation indices are fixed may be stored and accessed using the methods of Sec. 8.3.

This page intentionally left blank

Chapter 9

Indices

A derivative trade is a contractual agreement to exchange cashflows or other securities in the future, based on observed events in the market. To begin understanding the common features of such trades, we imagine ourselves in the far future, working out[1] the payments to which each party is contractually bound.

Through this thought experiment, we see a clear distinction between the market events – which enter the public record as *historical fixings* – and the terms of the trade, which describe the computation of a payout based on these fixings. Such historical fixings are not one-time events, but periodic (usually daily) snapshots of some ongoing process: an *index*.

Indices, defined in this way, have two crucial properties. First, they stand at the interface – to a large extent, they *define* the interface – between trades and models. Once a model can simulate the dynamics of indices, it need give no further consideration to trades.

Second, indices are the bridge connecting past and future. The computation of a payout from a set of index fixings is defined by the termsheet of a trade, and we will perform the same steps whether we are testing a scenario of the future, performing a postmortem on the past, or pricing and administering a live trade.

9.1   Naming and Parsing

An index's name plus the fixing date (or, on rare occasions, the fixing time) must completely describe its value. Thus the index name carries a lot of information, and we must work to define it with sufficient brevity.

---

[1] Amicably, of course.

The first task is to define a system of *canonical names* which provide an unambiguous representation of any index. These must be readily human-readable and also easily parsed.

We will sketch one such system, but it should not be regarded as definitive. Note that one goal is to restrict the number of special characters which must be reserved away from underlying names (*e.g.*, if _ is a special character to the parser, it is difficult to use it as part of an underlying).

We begin each canonical name with a mnemonic for the asset class, then the bracketed name of the underlying: EQ[IBM] or FX[USD/JPY]. Next we will have a separated list of increasingly-precise descriptors of the contract described by the index; as usual, our favored separator is the colon. Some indices will require an argument list (see Sec. 9.1.2); we use only named arguments and enclose the list in double-brackets. For time-shifted indices (forecasts of later values, or traded futures on an index) we use a special character (@ for a forward date, and > for a tenor or number of contracts to roll forward) immediately after the separator, followed by the date or tenor itself. Finally, fixing identifiers (nonfinancial information describing the index source, *e.g.*, identifying a particular Reuters page to use) will be appended at the end, bracketed and prefixed with FIX:. For example:

- EQ[IBM] – spot price of IBM.
- IR[USD]SWAP:LIBOR:5Y – five-year USD Libor-to-fixed swap with standard terms.
- IR[USD]SWAP:LIBOR:5Y[[FixedDayBasis=ACT/365L]] – as above but with some nonstandard terms.
- IR[USD]SWAP01:LIBOR:5Y[[FixedDayBasis=ACT/365L]] – the sensitivity of swap PV to coupon rate.
- IR[EUR]FUTURE:LIBOR@2014-09-17 – a September Euribor future.
- IR[EUR]FUTURE:LIBOR@2014-09 – a good parser will not need to be told the day.
- CM[WTI]FUTURE:>1 – the front WTI future. Note we use 1-offset.
- CM[WTI]FUTURE:>5bd>2 – the WTI future that will be the next-to-front contract in five business days. Note that this relies on our ability to associate WTI with NY holidays.
- CR[BCIT_SEN_DISCOUNT]CDS:5Y – a CDS rate with standard terms.
- IV[IR[USD]SWAP:LIBOR:5Y]@2014-09-17 – an implied Black vol to a fixed expiry date; see Sec. 9.4.
- IV[IR[USD]SWAP:LIBOR:5Y]>5Y[[Normal]] – the 5-into-5 swaption vol in normal rather than lognormal terms.

The standardized beginning of each canonical index string supports two important goals. First, it allows models to rapidly inspect indices when deciding what subset of the model is needed for pricing; see Sec. 10.0.1. Second, it allows us to replace a monolithic (and non-extensible) index parser with a singleton map of asset class tags to simpler parsers. Thus the top-level parser is essentially a switch on the leading type:

```
                        IndexParse.cpp
    // does not handle composite indices; see below
    Index_* Index::Parse(const String_& name)
    {
        int stop = name.find_first_of(":[");
5       if (stop == String_::npos)
        {
            return ParseSuperShort(name);
        }
        const String_ ac = name.substr(0, stop);
10      auto pp = TheIndexParsers().find(ac);
        REQUIRE0(pp != TheIndexParsers().end(),
                "No parser for '" + name + "'");
        return (*pp->second)(name);
    }
```

This demands an index parser class with a **Parse** member, and a registration function to populate **TheIndexParsers** at load time. We will further expand this parser in Sec. 9.2.1.

9.1.1 *Short Names*

Common indices, such as Libor swaps, may merit a "short name" which is recognized by the parser but different from the canonical name of the resulting index. The above approach supports two ways to handle short names. Those tagged with an asset class (such as IR:USD5Y) will be forwarded to the parser for that asset class; while "super-short" names containing no brackets or separator (such as USD/JPY) have their own (hopefully simple) parser.

9.1.2 *Nonstandard Indices*

If an index differs from the expected standard, we must decide whether the trade or the model should handle the aberration. For example, if a Libor swap has an unusual coupon frequency, the trade may form an odd kind of request to the model (we show an example of this above); or the trade may synthesize the swap itself from Libor forecasts and discount factors. The former approach is preferable if models support it; but eventually we will reach the limits of such support, and require trades to make this computation themselves.

The model's capabilities should reflect those implicitly assumed by term sheets (the contracts agreeing to a trade). If a term sheet can refer to an index by describing it, as opposed to defining its calculation method from other indices, then our trades should be similarly able to describe it to the model.

As our library matures, the range of supported indices will expand; then some trades can be simplified by rebooking them to take advantage of the new capability.

9.2   Fixings

Recall from Sec. 3.7.4 that we package access to in-process objects into the environment. We prefer to use the same method to access stored historical fixings.

The available historical fixings for a given index will form a single `Storable_` object, named after the index:

```
                         ———— Fixings.h ————
    class Fixings_ : public Storable_
    {
    public:
        typedef map<Time_, double> vals_t;
  5     const vals_t vals_;

        Fixings_
            (const String_& index_name,
             const vals_t& vals = vals_t())
 10        :
        Storable_("Fixings", index_name), vals_(vals) {}
    };
```

The functionality is intrinsically simple, and there is no need to dress it up in accessors.

We will wrap the task of checking for repository access in the environment, finding the appropriate `Fixings_`, and then testing for the presence of a desired element into a utility function:

```
                         ———— Index.cpp ————
    double PastFixing(_ENV, const String_& index_name,
```

```
                const Time_& time, bool quiet = false)
     {
         static const map<Time_, double> EMPTY;
   5     NOTE(index_name);
         const Fixings_* fixings = Environment::Find<Fixings_>
                 (_env, "No fixings exist", quiet);
         NOTICE("fixing time", time);

  10     auto vals = fixings ? fixings->vals_ : EMPTY;
         auto pf = vals.find(time);
         if (pf == vals.end())
         {
             Require(_env, quiet, "No fixing for this time");
  15         return DA::NAN;
         }
         return pf->second;
     }
```

──────────── *Index.cpp* ────────────
```
double Index_::Fixing(_ENV, const Time_& time) const
{
    return PastFixing(_env, Name(), time);
}
```

This does not fully address the problem of finding a fixing for an index. The issue is the treatment of FX fixings, which can be stored in two ways; e.g., as USD/JPY or as JPY/USD. Either of these is sufficient to define the other. We cannot store them under the same canonical name, because they have the same *information* but different *values* – one is about 10,000 times larger. We do not wish to store additional information in the index, because we want the canonical name to be a complete descriptor. Thus we must override the Fixing method for FX indices, which therefore needs to be a virtual method in the base Index_ class.

IndexFx.cpp
```
double Index::Fx_::Fixing(_ENV, const Time_& time) const
{
    const double test = PastFixing
            (_env, XName(false), time, true);
  5 return test != DA::NAN
            ? test
            : 1.0 / PastFixing(_env, XName(true), time);
}
```

This is not particularly efficient, but it is unlikely to be called multiple times except for complex path-dependent trades, which are not particularly rapid anyway. The utility function `XName` has the signature

```
───────────────── IndexFx.cpp ─────────────────
String_ Index::Fx_::XName(bool invert) const
```

and of course `Name` calls `XName(false)`.

This leads us to a definition of `Index_`:

```
───────────────── Index.h ─────────────────
class Index_ : noncopyable
{
public:
    virtual ~Index_();
5   virtual double Fixing(_ENV, const Time_& time_)
        const;   // calls Name
    virtual String_ Name() const = 0;
};
```

9.2.1   *Composites*

We can implement the functionality of `Index_` for a linear combination of indices. This seems baroque but allows a far better description of trades like "spread range accruals", which pay a coupon proportional to the number of days in which a spread between two indices is within some contractual limits.

```
───────────────── IndexComposite.h ─────────────────
namespace Index
{
    class Composite_ : public Index_
    {
5   public:
        typedef pair<Handle_<Index_>, double> component_t;
    private:
        Vector_<component_t> components_;
        double Fixing(_ENV, const Time_& time) const;
10      String_ Name() const;
    };
}
```

The implementation of `Fixing` is obvious, while that of `Name` must produce a syntax we can parse. We will support addition, subtraction, and left-multiplication by a constant; unary negation is optional. Thus "*single-index - single-index*" or "*2 * single-index + 3 * single-index*" should both be legitimate.

In defining the grammar, we can require that the single index names within a composite be bracketed, thus avoiding constraints on the syntax of single indices; or we can require that single index names never contain unbracketed arithmetic operators, thus ensuring that resolutions into such names (like the subtraction example above) will never break up a name. We prefer the former approach, which provides a friendlier user interface.

We now rename `Index::Parse`, from Sec. 9.1, to `ParseSingle`, and create a new function

IndexParse.cpp

```
Index_* Index::Parse(const String_& name)
{
    if (Composite_* test = ParseComposite(name))
        return test;
    return ParseSingle(name);
}
```

Here `ParseComposite` is a local function which splits on arithmetic operators (`+`, `-` and `*`) and calls `Index::ParseSingle` as necessary. There is no provision here for nested arithmetic operations, which have no practical value.

Note that, for purposes of storing a set of fixings, we should interpret indices with `ParseSingle`, since it makes no sense to store a composite index's fixing. Also, `ParseComposite` need not consider fixing identifiers, since the arithmetic that generates the composite fixing is not subject to such specializations; accordingly, the implementation of `Fixing` should be very simple.

9.3 Sorting and Hashing

We will see later (*e.g.*, in Sec. 10.2) the need for maps with `Index_` keys. If we use `Handle_<Index_>` as the key type, we must implement

```
bool operator<(const Handle_<Index_>& lhs,
       const Handle_<Index_>& rhs)
{   // deprecated
    return lhs->Name() < rhs->Name();
}
```

to support such maps. No other implementation is readily available, since we cannot enumerate in advance the types of indices.[2]

This is computationally expensive because `Name` must be evaluated twice for each comparison, and many more for a binary search or map insertion. We can do better without much implementation effort:

```
                                     ─── Index.h ───
    struct IndexKey_
    {   // constructor depends on member order!
        const String_ name_;
        const Handle_<Index_> val_;
5
        // allow empty handle, but not implicitly
        IndexKey_(const Handle_<Index_>& val)
           : val_(val),
        name_(val.Empty() ? String_() : val->Name())
10      {    }

        IndexKey_(const Index_& val)
           :
        name_(val.Name()),
15      val_(Index::Parse(name_))
        {    }

        const Index_* operator->() const {return val_.get();}
    };
20  inline bool operator<
        (const IndexKey_& lhs, const IndexKey_& rhs)
    { return lhs.name_ < rhs.name_; }
    inline bool operator==
        (const IndexKey_& lhs, const IndexKey_& rhs)
25  { return lhs.name_ == rhs.name_; }
```

This can be used in place of `Handle_<Index_>` when an associative array uses indices as keys. The same approach supports hash maps; we will hash the precomputed `name_` as needed.

9.4 Implied Vol

In some highly structured products, we are called on to observe the implied volatility of some underlying index. This implied vol has all the properties

---

[2]One could imagine comparing the vtable addresses of indices, thus saving a string comparison for indices of different derived types, but this would be depraved.

of an index; so it is itself an index which requires parsing, and we will store the underlying index within the implied vol index.

When searching for historical fixings, the implied vol is treated like any other index. During model valuation, the model must be able to compute implied volatilities and also have knowledge of the underlying index; thus the latter is part of our new index's public interface.

```
————————————————— IndexIv.h ——————————————
class IndexIv_ : public Index_
{
   Handle_<Index_> underlying_;    // no trade/fixing ID
   Cell_ expiry;         // date or maturity tenor
5  Maybe_<double> callDelta_;    // otherwise ATM
   VolType_ volType_;
public:
   String_ Name() const;
   // help models find what is expected
10 const Index_& Underlying() const {return *underlying_;}
   Maybe_<double> CallDelta(bool invert = false) const;
   long ExpiryDate(long fixing_date) const;
   const VolType_& VolType() const {return volType_;}

15 IndexIv_
      (const Handle_<Index_>& underlying,
       const VolType_& vol_type,
       long expiry_date,
       const double* call_delta = 0);
20 IndexIv_
      (const Handle_<Index_>& underlying,
       const VolType_& vol_type,
       const String_& expiry_tenor,
       const double* call_delta = 0);
25 };
```

This page intentionally left blank

Chapter 10

Pricing Protocols

We have displayed an array of tools underpinning the creation of trades, forward curves, models and numerical pricers. To put these pieces to use, we have to define the target. A trade, for example, will be useful if it does certain trade-like things: but what precisely are those?

In answering this question, we are creating protocols for communication between high-level types. Regrettably, these protocols are usually not made explicit or given any serious design effort. There is no avoiding the existence of protocols — except by wholly sacrificing reuse — but without an active effort they will be implicit, haphazard and scattered amongst unrelated code.

Since our central task is pricing, we begin by considering the information flow during a numerical pricing task. (Closed-form pricing, which often requires a much more intimate communication between trade and model, is discussed in Ch. 14.) Bear in mind that the model should in no way be specific to the trade: in particular, it may describe underlyings which are not relevant to the trade or may parametrize curves far beyond the trade's maturity. Thus the trade must begin by communicating its underlyings, and the model must respond by specializing to these. Recall from Sec. 2.3 that this is not accomplished by the model's changing itself, but by producing a new object of a different type. At this point we also commit to a numeraire currency, or *value currency*; this can be an input to the pricing request or an attribute of the trade. We prefer the latter; in the unusual event that we need to change value currencies, we can easily do so by embedding the trade in a composite (see Sec. 11.8).

Thus the model is responsible for three things: it must describe the evolution of its internal state variables, compute index values from the state variables at a node, and compute discount factors for whatever payments arise from the trade. For the first task, the model creates *steppers*, each of which (for a Monte Carlo) takes a step based on input random deviates and (for a PDE) computes coefficients as in Sec. 7.9. For the others, the model produces a *value request* to be sent to the trade – this is like a form for the trade to fill out to state what index values it will need to see, and what payments it may make.

The value request makes *promises* to the trade about where the needed values may be looked up in the future. The trade combines these promises to create a *payout* – in fact, the request is filled out in the process of creating the payout. The event times (on which indices are observed) are determined from the request, and used to form the necessary steppers. The model then creates an *asset* which honors the promises made to the trade by the request. Similar promises are made for discount factors required for the trade's payments; these are not visible to the trade but the model handles them identically.

For backward induction (*i.e.*, PDE) pricing, this is all the model can do: the trade will manipulate values in the PDE grid, and the numerical engine will roll them back, until we reach the present.

Monte Carlo pricing allows additional control over the process, and also introduces other sources of complexity.

- The payout may be path-dependent, *i.e.*, it may have a *state* of its own.
- The model may present events – such as defaults – which are not bound to the trade's event times.
- Bermudan exercise is no longer easily handled by the payout.
- The trade may depend on the daily path of indices, as for range accrual trades.

The last of these can be avoided, if we drop the idea of holistically representing the path of an index; *e.g.*, we would think of a range accrual trade as having a trade event on every business day. However, a whole-path representation – an *index path* – is a crucial performance improvement for such trades, and is also very useful in other contexts.

To deal with Bermudan exercise decisions, the payout must describe its exercises in the form of *backward induction actions* which can be taken by the numerical pricer after all paths have been run.

10.0.1 *Which is a Model?*

Even in the heuristic discussion above, we have risked confusion by using the word "model" in conflicting ways. There are two distinct roles here. The first is an object that defines a function mapping trades to values; this would properly be called a "pricer", but "model" is the established nomenclature. The "model" may lack any consistent dynamics (as in a swap market model) or may recalibrate internal parameters in a trade-dependent way; but it is invariably still called a model, however little it may deserve the appellation.

The second use is an object that defines the evolution through time of market observables. This is a model in the mathematical sense; but that name is already taken. "Pricing model" sounds too much like pricers (*i.e.*, models) to be safely used. In this work we use "SDE", which is not strictly accurate but does convey the sense of a process (or set of processes) in a particular measure. Thus the model creates an SDE based on the trade's underlying and value currency; and the SDE in turn creates the steppers, value request, and asset. We will still sometimes refer loosely to "the model" when there is no prospect for confusion, but it should be understood that it is the SDE which interacts directly with the numerical pricing engines.

10.1 Past and Future

The above discussion is also complicated by the necessity to distinguish somewhere between "past" and "future" – between events that are now historical and those that are still uncertain. We distinguish several ways of making this distinction, each defined by its own "present":

- *Reset time* – after which index fixings after unknown.
- *Accounting date* – before which payments have "rolled off" and are ignored.
- *Value date* – on which the discount factor is 1.
- *Vol start time* – at which stochastic processes "turn on."

As a rule, all these will be the same; however, when computing various decay scenarios we may change them in any order.

The vol start time is an attribute of the model; the reset time and accounting date are parameters of the valuation, or in the environment. The value date can be treated either way; on balance we find it better to treat it as a parameter of the valuation.

The reset time is the most fundamental of these, since it determines the meaning of a request for an index fixing; it partitions past from future in all our numerical methods.

It is important to construct the valuation process so that the trade does not see this partition: this avoids substantial code duplication in each concrete trade type. The model will be less protected, because its definition of "present" (the vol start time) may not coincide with the reset time. We will mediate the trade's requests to ensure that the model is not responsible for lookup of historical data.

10.2   Underlyings

The underlying is meant to support the creation of an appropriate SDE, so it provides fairly restricted information.[1]

```
─────────────────── Underlying.h ───────────────────
struct Underlying_
{
    map<String_, long> payCcys_;
    map<IndexKey_, Time_> indices_;
5   map<String_, long> credits_;

    Underlying_& operator+=(const Underlying_& more);
};
```

For each currency paid, we note the last pay date; for each index, the last event time at which it is observed; and for each reference credit, the last default date on which it need be monitored.

An alternative is to have the underlying provide complete information about the usage times of every index, which could then be used in place of the value request. However, this is simply too much information for many purposes; thus we prefer the simpler underlying.

10.3   Payments and Streams

So far we have not discussed how the cashflows generated by the trade are to be discounted. We generally divide this discounting into two portions: a factor $D(t)$ from today to the event time, and a zero-coupon bond price

---

[1]We may extend the underlying slightly to support trade-specific calibration: see Sec. 14.7.

$P(t, T)$ at the event time. The trade could extract the latter itself, since both $P(t, T)$ and any necessary FX rate can be represented as index fixings and extracted through the value request; but this would be a mistake.

The problem is that the discounting is absorbed into the opaque workings of the trade, and commingled with the computation of amounts. Part of our job is to support the ongoing maintenance of real trades, not just valuation; and a report of the payments to be made is an important part of this maintenance.

Thus, rather than discount its own payments, we will force a trade to register a possible payment, obtaining for each a *payment tag* which will mediate the discounting during valuation. At first glance this might seem like sheer obfuscation; but it lets us use the same machinery for valuation and for other kinds of reporting (e.g., a list of expected upcoming payments). Thus it saves us from duplicative and error-prone recoding in each concrete trade type.

A payment tag is not much to look at:

—————————————— Payment.h ——————————————

```
namespace Payment
{
    class Tag_ : noncopyable
    {
    public:
        virtual ~Tag_();
    };
    const Handle_<Tag_>& Null();

    namespace Amount
    {
        class Tag_ : noncopyable
        {
        public:
            virtual ~Tag_();
        };
    }
}
```

Null here is the equivalent of the Unix /dev/null – it is a tag for payments that do not matter (generally because they precede the accounting date). Making it visible in this way allows this fact to be communicated to the trade, allowing some unimportant optimizations.

We make a distinction in code between payments, which have financial value and may be discounted, converted to different currencies, and so on;

and raw amounts, which are just numbers. This lets us write clearer and safer code with no extra run-time cost.

10.3.1 *Payment Reporting*

More interesting is the information which the trade provides in order to obtain the tag. This (also in `namespace Payment`) gives some description of the payment as well as the minimal information to compute the discount factor. The descriptive information is ignored during valuation but makes the payment report more useful.

```
                          ──────── Payment.h ────────
   namespace Payment
   {
      struct Conditions_
      {
 5       enum EXERCISE
         {
            UNCONDITIONAL,
            ON_EXERCISE,
            ON_BARRIER_HIT,
10          ON_CONTINUATION
         } exerciseCondition_;
         enum CREDIT
         {
            RISKLESS,
15          ON_SURVIVAL,
            ON_DEFAULT
         } creditCondition_;
         // if paid on default, we need still more info:
         DefaultPeriod_ defaultPeriod_;
20
         Conditions_();   // unconditional, riskless case
      };

      struct Info_
25    {
         String_ description_;
         Time_ knownTime_;
         Conditions_ conditions_;
         Maybe_<AccrualPeriod_> period_;
30       Info_(const String_& des = String_(),
               const Time_& known = Time::Minimum(),
               const Conditions_& cond = Conditions_(),
               const AccrualPeriod_* accrual = 0);
```

```
35      };
   }

   struct Payment_
   {
       Time_ eventTime_;
40     String_ ccy_;
       long date_;
       String_ stream_;
       Payment::Info_ tag_;
       long commitDate_;
45     Payment_();    // support Vector_<Payment_>
       Payment_(const Payment_& src);
       Payment_(const Time_& et, const String_& ccy, long dt,
               const String_& s, const Payment::Info_& tag,
               long cd = Date::Minimum());
50 };
```

The struct `Conditions_` is an indication of the kind of information that might be included in a payment report. We nest `Info_` inside `namespace Payment`, rather than inside `struct Payment_`, so that it can be forward-declared in other headers.

10.3.2 *Commitment to Streams*

This introduces the concept of a *stream*, into which payments are directed. For those familiar with PDE pricing, a stream is essentially the thing whose value is rolled back on the PDE grid: it is a collection of payments which can be summed for purposes of valuation. The two legs of an ordinary swap can be collected into a single stream. A Bermudan swaption has two streams (the swap, and the option to enter into the swap) though we might in practice roll back only the latter and price the former *ab initio* at each node. In either case, the trade value is obtained from the stream values by some linear relation, which the trade must specify.

A bond, for instance, is a stream. If we take possession of the bond at a specified *delivery date* – due to an option exercise, for instance – we receive the coupons whose payment dates are after delivery. This is represented by the `commitDate_` above, which for a bond coupon is just the coupon payment date; it is the date which is compared to the delivery date to determine whether the payment is received. We will represent the accrued interest in a bond option as a fee paid upon exercise.

There is one complication: truly American options, or Bermudan callable swaps with an exercise frequency higher than the coupon frequency. For the latter, upon exercise we must pay accrued interest for the elapsed part of a (possibly floating) period. We can deal with this by complicating the framework, extending the `commitDate_` above to an interval (start date, end date, and day basis) which will support fractionally-committed payments like this one; or by complicating the trade, breaking the coupon payments into several payments with different commit dates (but otherwise identical). The decision depends on the business mix we are expecting; here we will continue with the simpler framework.

10.3.3 *Destinations*

The precise form of the payout depends on the method for committing payments to streams. For this we use a simple abstract class, with a payment represented by a call to `operator+=`:

```
───────────── Payment.h ─────────────
class NodeValue_ : noncopyable
{
public:
    virtual ~NodeValue_();
5   virtual void operator+=(double amount) = 0;
    // direct support for backward induction:
    // virtual double& operator*() = 0;
};

10 class NodeValues_ : noncopyable
{
public:
    virtual ~NodeValues_();
    virtual NodeValue_& operator[]
15      (const Payment::Tag_& tag) = 0;
    inline NodeValue_& operator[]
        (const Handle_<Payment::Tag_>& tag)
    { return operator[](*tag); }

20  virtual double& operator[]
        (const Payment::Amount::Tag_& tag) = 0;
    inline double& operator[]
        (const Handle_<Payment::Amount::Tag_>& tag)
    { return operator[](*tag); }
25 };
```

The `operator*` in `NodeValue_` requires some explanation. In Monte Carlo valuation, we simply add payments to the streams; manipulation of stream values (for example, to reflect Bermudan exercise) is not meaningful within a single path and requires separate machinery (see Sec. 10.7 on backward induction actions, and Sec. 7.10). In PDE valuation, we have two main alternatives: we can allow the trade to directly manipulate the stream values (as `NodeValue_::operator*` does), or we can require the PDE to apply the same backward induction actions formed for the Monte Carlo.

The decision whether to supply this function will be based on our opinion of the importance of PDE pricing: does it merit direct support by trades in order to simplify the PDE implementation? My own preference is to use PDE pricing very little, instead favoring large, flexible models whose high dimensionality necessitates Monte Carlo pricing. In this work, I will not show PDE-specific code.

In forming the backward induction actions, we set amounts (never payments!) for them to use, *e.g.*, as American Monte Carlo observables.

10.4 Index Paths

We have discussed the desirability of accessing the whole path of an index, not just discrete fixings, in computing the payout. This is a major commitment for models – once we take this route, models whose SDE's cannot produce index paths will fail to price trades which attempt to use them.

```
————————————— IndexPath.h —————————————
class IndexPath_ : noncopyable
{
public:
    virtual ~IndexPath_();

5
    virtual double Expectation
        (const Time_& fixing_time,
         const pair<double, double>& collar)
    const = 0;

10
    virtual double FixInRangeProb
        (const Time_& fixing_time,
         const pair<double, double>& range,
         double ramp_width = 0.0)
15  const = 0;
```

```
    // Probability of staying strictly inside a range
    virtual double AllInRangeProb
      (const Time_& from,
       const Time_& to,
       const pair<double, double>& range,
       double monitoring_interval,
       double ramp_sigma = 0.0)
    const = 0;

    virtual double Extremum
      (bool maximum,
       const Time_& from,
       const Time_& to,
       double monitoring_interval,
       const pair<double, double>& collar)
    const;    // default calls AllInRangeProb
};
```

The output of each function is an expectation conditional on all state information at event times; thus use of the index path is essentially path integration over the "blank space" between event times. It follows that a linear combination of these, such as a number of days in range, is also a conditional expectation. This lets us mix a single piece of whole-path information into the payout with complete freedom; and conditionally independent quantities (*e.g.*, days in range over disjoint intervals separated by an event time) can also be used freely. Conversely, nonlinear functions of whole-path information would not be priced correctly by this method; in the unlikely event that one should become important, we would have to introduce a new member in `IndexPath_` to evaluate it.

The implementation of the path will likely use Brownian-bridge methods, which we work to make accurate but which cannot be perfect. The errors thus introduced can be measured, and if necessary controlled, by inserting fake event times into the payout – this will more tightly constrain any Brownian interpolation and approach the limit of using only information at event times.

10.4.1 *Historical Paths*

A path which is completely in the past – *i.e.*, all its fixings are before the reset time – clearly should have a model-independent implementation. Also, such a path might need to be part of a model-constructed `IndexPath_`;

e.g., the times in a call to **Extremum** might span the reset time. We create a transparent data structure for this purpose:

```
struct IndexPathHistorical_ : IndexPath_
{
    map<Time_, double> fixings_;
    double Expectation
        (const Time_& t,
          const pair<double, double>& lh)
    const
    {
        auto pFix = fixings_.find(t);
        REQUIRE0(pFix != fixings_.end(),
            "No fixing exists for " + String::FromTime(t));
        return Max(lh.first, Min(lh.second, pFix->second));
    }
    // ...
```

The implementations of the other member functions of **IndexPath_** are equally simple. In the unusual case of trades that depend on high-frequency tick data, we may need to reimplement **IndexPathHistorical_** to store open-high-low-close information.

10.5 Defaults and Contingent Payments

We may need to price trades which depend on the defaults in some pool of reference securities; thus the presence of **Underlying_::credits_** above. Should the model simulate a default event, this must be communicated to the trade and to the numerical method, which are concerned with different aspects of the event. Thus we first isolate one part for the trade to see:

Default.h
```
struct ObservedDefault_
{
    long date_;
    CreditId_ referenceName_;
    double recovery_;
};
```

Then we write, in another header not seen by trades,

DefaultNumerical.h
```
class DefaultEvent_ : noncopyable
{
protected:
```

```
         DefaultEvent_(long date, const CreditId_ which,
 5           double recovery, double df_from_event);
     public:
       virtual ~DefaultEvent_();

       const ObservedDefault_ observed_;    // seen by trade
10     const double dfFromPreviousEvent_;    // seen by MC
     };
```

Subclasses of `DefaultEvent_` may contain other information seen only by
the SDE and its steppers; the trade and Monte Carlo will not care.

The `CreditId_` is a proxy for the reference name of the defaulted credit.
Multi-credit trades will almost surely need to look up this identifier in a
sizeable table of names; *e.g.*, a map of names to notionals for a CDO.
We introduce `CreditId_` to prevent a large number of string comparisons,
which can add significantly to the running time. Its role is simply to support
rapid comparison:

———————— *Default.h* ————————
```
struct CreditId_
{
    int val_;
    explicit CreditId_(int p) : val_(p) {}
 5 };
inline bool operator<(const CreditId_& lhs,
        const CreditId_& rhs) {return lhs.val_ < rhs.val_;}
```

The id's must remain valid for the duration of the pricing. In practice, we
accomplish this by storing the reference credits in an object that will not be
destroyed before pricing is complete, then doling out their addresses within
that object.

———————— *Default.h* ————————
```
struct AssignCreditId_ : noncopyable,
        unary_function<String_, CreditId_>
{
    Vector_<String_> names_;    // constructor sorts
 5  AssignCreditId_(const Vector_<String_>& names);
    CreditId_ operator()(const String_& name);
};

CreditId_ AssignCreditId_::operator()(const String_& nm)
10 {
    auto pn = LowerBound(names_, nm);
    REQUIRE0(pn != names_.end(), "Reference credit '" + nm
```

```
         + "' is not part of the trade underlying");
      return CreditId_(pn - names_.begin());
15 }
```

The pricing routine will create a single instance of this object, based on the trade's stated underlyings, to be shared by the trade and SDE during the setup phase. Its mission is then accomplished.

10.5.1 *Immediate Payments*

Ideally, the trade (*i.e.*, the payout) would receive the default information, update its state accordingly, and then make payments at the next event time. However, a trade might generate cashflows immediately upon default, rather than politely waiting for an event time, and we must prepare for that contingency. Thus we need a destination, akin to `NodeValues_`, for such immediate payments:

```
 _____ Payment.h _____
class NodeValuesDefault_ : noncopyable
{
public:
    virtual ~NodeValuesDefault_();
5    virtual NodeValue_& operator()
       (const Payment::Default::Tag_& tag,
        long commit_date) = 0;
    inline NodeValue_& operator()
       (const Handle_<Payment::Default::Tag_>& tag,
10        long commit_date)
    { return operator()(*tag, commit_date); }
};
```

Here the tag (now of type `Payment::Default::Tag_`, but just as blank and uninteresting as `Payment::Tag_`) is determined by the stream into which the payment is made. Since the commitment date cannot be determined without knowing the default date, it is left for the trade to supply along the path.

10.5.2 *Viewing Indices*

The payment might not be truly immediate: more commonly, it takes place some days or months after the default event. Who should be responsible for discounting the payment? It turns out that this is much more simply done by the payout; the trade can describe the necessary discount rate as an `Index_`, and the payout can read its value from the `IndexPath_`.

This shows the desirability of making index paths visible to the payout when a default occurs; also, the paid amount might itself depend on market quantities at the time of default.

10.6 Requests and Promises

The form of the value request is dictated by the form of the promise it makes. This is some abstraction of an address: a location to which the needed value will later be written. Indeed, it can be a literal address, probably `volatile const double*` for values – this maximizes efficiency, while a less explicit promise will leave us more implementation freedom. In practice, the literal address makes multi-threading unnecessarily complicated, and we will avoid it. We would prefer to promise something compact and inscrutable, like an `int`, and delegate its interpretation completely to the asset.

```
───────────────── AssetValue.h ─────────────────
   namespace Valuation
   {
       typedef std::size_t address_t;
       Maybe_<double> KnownValue(address_t loc);
5      Maybe_<address_t> FixedLoc(double value);

       struct IndexAddress_
       {
           int val_;
10         IndexAddress_(int val) : val_(val) {}
       };
   }
```

But we can hide rather a lot inside a (32-bit) integer! One useful thing to hide is an indication of time or event index, which can be used inside the asset to ensure that the trade is not "peeking ahead" to part of the path which has not yet been formed. This is not part of the interface to trades, which receive an `address_t` already formed and later use it, unaltered, to obtain a value from the model.

We can also hide a set of fixed values inside an address: the trade can use `KnownValue` to see if the address points to a deterministic quantity, and the SDE can use `FixedLoc` to see whether a given number has a value.

(The latter function can contain a singleton registry of addresses and add to it as needed, but this is hostile to multi-threading.) This can sometimes support optimizations within the payout, though our preference is to use it only for 0.0 and 1.0.

Index addresses have their own type, and here **typedef** is not a strong enough distinction: two typedefs which resolve to the same type are interchangeable to the compiler. Thus we create a struct which simply wraps an **int**.

If there is no abstraction penalty in performance, we can do the same for value requests, replacing **address_t** with a **struct**. The best test for abstraction penalty is to change the code and see whether the built (non-debug) dll's change at all. Essentially any change represents a performance degradation.

The asset in its entirety will not be available to the payout; instead we will send a token, also defined in **namespace Valuation**:

```
─────────── AssetValue.h ───────────
class UpdateToken_
{
    typedef Vector_<Handle_<IndexPath_>> indices_t;
    Vector <>::const iterator begin ;
5   indices_t::const_iterator indexBegin_;
    const int valMask_, dateMask_;
public:
    const Time_ eventTime_;
    UpdateToken_(Vector_<>::const_iterator begin,
10           indices_t::const_iterator index_begin,
             int val_mask, int date_mask,
             const Time_& event_time);

    inline const double& operator[]
15      (const Valuation::address_t& loc) const
        {
        // loc&dateMask_ can be checked with assert
        return *(begin_ + (loc & valMask_));
        }
20   const IndexPath_& Index
        (const Valuation::IndexAddress_&) const;
};
```

In a better world, **UpdateToken_** would be an abstract class with a virtual **operator[]**; but here, at the hottest of all code hotspots, we begrudge even the overhead of an arithmetic operation, and cannot tolerate a virtual function call.

The `eventTime_` must be part of the token's public interface, so that payouts can look at the token to determine what actions to take.

Now, who will make these promises to the trade?

```
─────────── ValueRequest.h ───────────
class ValueRequest_
{
public:
    virtual ~ValueRequest_();

    virtual Handle_<Payment::Tag_> PayDst
        (const Payment_& flow) = 0;

    virtual Handle_<Payment::Default::Tag_> DefaultDst
        (const String_& stream) = 0;

    virtual Valuation::address_t Fixing
        (const Time_& event_time,
         const Index_& index) = 0;

    virtual Valuation::IndexAddress_ IndexPath
        (const Time_& last_event_time,
         const Index_& index) = 0;

    typedef Valuation::address_t address_t;
    typedef Valuation::IndexAddress_ IndexAddress_;
};
```

This will give the trade all it needs to form a payout, which will in turn be able to compute the cashflows at each node once given the appropriate `UpdateToken_`.

10.6.1 *Help for Models*

In a separate header, not visible to trades, we expand this interface slightly:

```
─────────── ValueRequestImp.h ───────────
class IndexPathHistory_ : noncopyable
{
public:
    virtual ~IndexPathHistory_();
    virtual Time_ ResetTime() const = 0;
    virtual Handle_<IndexPathHistorical_> History
        (const Index_& index)
    const = 0;
```

```
};
class ValueRequestImp_ : public ValueRequest_
{
public:
    virtual address_t InsertFixing(double amount) = 0;

    virtual IndexAddress_ IndexPath
        (const Time_& last_event_time,
         const Index_& index,
         const IndexPathHistory_* historical) = 0;
    IndexAddress_ IndexPath
        (const Time_& last_event_time,
         const Index_& index)
    {
        return IndexPath(last_event_time, index, 0);
    }

    virtual IndexAddress_ InsertPath
        (const Handle_<IndexPath_>& path) = 0;

    virtual map<IndexKey_, address_t> AtTime
        (const Time_& t) const = 0;
};
```

As discussed in Sec. 10.1, the `ValueRequest_` must distinguish between requests before and after the reset time. In practice, the SDE will form a `ValueRequest_` object, which the numerical method will then decorate (see Sec. 2.6) to pass to the trade. The `InsertFixing` and `InsertPath` functions support this separation; see Sec. 10.13.6. The `AtTime` function allows the `SDE_` to discover what promises have been made by a `ValueRequest_`; see Sec. 13.3.

10.6.2 *Destinations*

In PDE valuation, we will be accumulating a separate value for each stream (thus our PDE solver, in Sec. 7.9, handles a vector of payouts). The output of `PayDst` is an abstraction of the index into this vector; in Monte Carlo valuation, where accumulation often must be deferred, it is a more detailed label. But the trade knows nothing of this: it simply treats the tag as a kind of subscript.

Payments before the accounting date can be ignored; we accomplish this by returning `Payment::Null()` from the `ValueRequest_`. The trade might check the returned handle (by pointer equality) and optimize accordingly (by not computing the amount at all), but in practice this is not an important optimization.

10.7   Bermudans and Barriers

A European option can be priced using Monte Carlo in two ways: as a cap or floor (often zero) on the payment amount into a single stream, or as a conditional transition from one stream (in which no payment is made) to another (where the option's underlying is paid). The former approach is simpler, especially for trades containing several independent options (*e.g.*, interest rate caps or cliquets). The latter is necessary, however, for Bermudan exercises when there are several opportunities to make the transition between streams (or even for Europeans when the underlying itself requires Monte Carlo valuation).

Before we can define an option, we must be able to define what is received upon exercise. This is a combination of single payments (which we refer to as *fees*, regardless of their sign) and portions of streams. First we define these portions:

```
                         ── BackwardInduction.h ──
namespace BackwardInduction
{
    using Payment::Tag_;    // for fees
    typedef Payment::Amount::Tag_ amount_t;
5
    struct StreamSegment_
    {
        String_ stream_;
        long deliveryDate_, terminationDate_;
10      StreamSegment_(const String_& s, long td,
            long tt = Date::Maximum());
        StreamSegment_();    // support vectors
    };
```

Given this method for describing the underlying, we form different actions to take during backward induction. For example, to define the result of a Bermudan exercise decision, we must know not only the stream and delivery date but also the sign (long or short) of the option, the fees paid and streams received on exercise (if any), and the observables (see Sec. 7.10).

```
────────────────── BackwardInduction.h ──────────────────
     // Different concrete actions
     struct Exercise_
     {
         int sign_;
5        // where to find the values
         vector<Handle_<Tag_> > fees_;
         vector<StreamSegment_> underlyings_;
         vector<Handle_<amount_t> > observables_;
         vector<Handle_<amount_t> > slaves_;    // write exProb
10   };
```

We also prefer to think of barrier hitting, in a knock-in or knock-out option, as a stream transition. It is equally possible to think of it as a modification to the payout state, which then affects all later payments; but making the stream transitions explicit has substantial operational value. There may be payments if the barrier is hit, or streams which knock in; and we must know the probability of hitting.

```
────────────────── BackwardInduction.h ──────────────────
     struct Barrier_
     {
         vector<Handle_<Tag_> > payOnHit_;
         vector<StreamSegment_> knockIn_;
5        Handle_<amount_t> hitProb_;
     };
```

This is also important for some applications, such as asset swaps on callable bonds, where it is necessary to make a change to one stream when a different stream is called. The former stream is said to be "slaved" to the latter. This is the purpose of Exercise_::slaves_; we create a Barrier_ action in the slaved stream, and add its hitProb_ member to the list of slaves_ of the exercise point in the master stream. Upon making the exercise decision, we write the result (0 or 1) to that location; it is then available for use by the slaved stream.

The HitProb_ is an amount, not a boolean flag, for several reasons. First, we already have a mechanism for transmitting values. Second, the value may be determined from an index path using AllInRangeProb, which can give values over the whole range $[0, 1]$. Finally, even when it is not, we will often artificially smooth the hit probability to give continuous and/or conservative values; continuous values are crucial for numerical hedge computation.

Another action, `Include_`, simply adds one stream's payments to another. This lets us write trades, should we choose, with finer granularity: we can represent legs as separate streams, then pull them together using `Include_`.

```
──────────────── BackwardInduction.h ────────────────
    struct Include_
    {
        vector<StreamSegment_> src_;
    };
```

Exactly one of the above specific actions is required to form a valid generic action. The proper high-tech solution is a `boost::variant`, which accomplishes exactly this. Since we must support vectors of actions, we need a default constructor which perforce contains no valid action; thus we add `boost::empty` as an alternative action type.

```
──────────────── BackwardInduction.h ────────────────
     struct Action_
     {
         String_ stream_;
         Time_ eventTime_;    // not strictly necessary
 5       long deliveryDate_;
         variant<Exercise_, Barrier_, Include_, empty>
             details_;
         Action_(const String_& stream,
             const Time_& event, long delivery);
10       Action_();   // support vector<Action_>
     };
}    // leave namespace BackwardInduction
```

The `Action_` is a passive **struct**, communicating data from the trade for use by the numerical pricing engines.

10.8 Payouts

At last we can define the payout of a trade.

```
──────────────── Payout.h ────────────────
class Payout_ : noncopyable
{
public:
    typedef Handle_<Payment::Tag_> dst_t;
 5  typedef Handle_<Payment::Amount::Tag_> amount_t;
    virtual ~Payout_();
```

```
     virtual Vector_<Time_> EventTimes() const = 0;

10   class State_ : noncopyable
     {
     public:
        virtual ~State_();
        virtual State_* Clone() const = 0;
15      virtual State_& operator=(const State_& rhs) = 0;
     };
     virtual State_* NewState() const;
     virtual void StartPath(State_* state) const;

20   virtual void DoNode
        (const Valuation::UpdateToken_& values,
         State_* state,
         NodeValues_& pay_dst)
     const = 0;

25
     virtual void DoDefault
        (const ObservedDefault_& event,
         State_* state,
         const NodeValuesDefault_& pay_dst)
30   const;    // base implements a no op

     virtual Vector_<BackwardInduction::Action_>
        BackwardSteps() const = 0;

35   // components of value for each trade name
     virtual map<String_, Vector_<pair<String_, double> > >
        StreamWeights() const = 0;
     };
```

The final member function, StreamWeights, provides the translation from stream values to trade values. It takes this complex form because a trade may have several payouts (Sec. 11.8 describes composite trades; we can even have customized trades involving more than one counterparty), each of which is a linear combination of stream values. The top-level trade names are the keys of the output of StreamWeights.

10.8.1 *Trade State*

Payout_::State_ is simply a placeholder to be carried by the Monte Carlo; each path-dependent trade will create its own subclass, which will be returned to it at each call of `DoNode` or `DoDefault`. A PDE, which naturally does not support path dependence, will pass a null pointer instead. Thus payouts must check that the state exists before using it, unless we add some other way to signal path dependence.

10.8.2 *Values Store*

The `NodeValues_` must translate from *amounts* of the cashflows made by the trade to *values* which can be discounted, converted to a common currency, and summed over payment times. To accomplish this, we will provide the `ValueRequest_` during construction of the `NodeValues_`, so that its demands for market information can be joined with the trade's.

10.9 Steps

In contrast to the machinery above, the interface between SDEs and numerical methods is quite simple. Partly this is due to the extreme opacity of models: they have state variables, which change stochastically through time; they use these state variables to generate observable fixings; and there ends the story.[2]

For the PDE, the interface of our solver (Sec. 7.9) determines what information the stepper must provide. The Monte Carlo interface is more opaque – since the dynamics are completely concealed – but richer because of additional possibilities:

- The steps may be path-dependent (*e.g.*, a credit step may depend on which credits have already defaulted), and this information should be communicated to the stepper.
- The steps may store a *paths record* with information about the run, which can be used to stabilize risk in bumped runs.
- The stepper must be fully bitwise constant (no `mutable` members) so that it can be shared by multiple threads.

We require a few supporting classes first:

```
―――――――――――――― MCPath.h ――――――――――――――
namespace MonteCarlo
{
    struct Workspace_ : noncopyable
```

---

[2]Our use of a `Vector_<>`, rather than a generic `struct`, to store model state is a concession to the PDE.

```
       {
  5       virtual ~Workspace_();
          virtual Workspace_* Clone() const = 0;
       };
       struct PathsRecord_ : noncopyable
       {
 10       virtual ~PathsRecord_();
          virtual void StartPath(int i_path) = 0;
       };
       typedef boost::shared_ptr<PathsRecord_> record_t;
     }
```

We put the paths record inside the workspace to pass to the stepper. The workspace will also contain, *e.g.*, temporary vector storage (to avoid allocating memory at each step; see Sec. 2.8).

```
  ─────────────────── Step.h ───────────────
  struct ModelStepper_ : noncopyable
  {
     virtual ~ModelStepper_();

  5  // PDE interface
     virtual PDE::ScalarCoeff_* DiscountCoeff() const;
     virtual PDE::VectorCoeff_* AdvectionCoeff() const;
     virtual PDE::MatrixCoeff_* DiffusionCoeff() const;

 10  // MC interface
     virtual MonteCarlo::Workspace_* NewWorkspace
        (const MonteCarlo::record_t& paths_record)
     const = 0;
     virtual int NumGaussians() const = 0;
 15  virtual void Step
        (Vector_<>::const_iterator iid_gaussian_begin,
        Vector_<>* state,
        MonteCarlo::Workspace_* work,
        Random_* more_randoms,
 20     double* rolling_df,
        vector<Handle_<DefaultEvent_> >* defaults)
     const = 0;
  };
```

The paths record must be shared across steps. Also, for many models the parameters used in a stepper depend on some quantity accumulated before the step start date. For example:

- In BGM-like models, we may linearize the drift, which can be done more accurately if we know the moments of the state at the step start time.
- In Markov chain models, jump compensator terms depend on the distribution of the Markovian state.

These and similar computations can in principle be performed *ab initio* for each stepper being created, but – especially in the presence of many short steps – it is far more efficient to maintain and update a cumulative state. Since we will use this object in the creation of steppers, it makes sense to associate the paths record with it as well:

Step.h

```
struct StepAccumulator_ : noncopyable
{
    virtual ~StepAccumulator_();

5   virtual MonteCarlo::PathsRecord_* NewPathsRecord
        (int num_paths,
         const MonteCarlo::record_t& base_record)
        const
        { return 0; }
10
    virtual Vector_<pair<double, double> > Envelope
        (const Time_& t,
         double num_sigma)
        const = 0;
15 };
```

Sometimes the stepper has a need for additional random variables – *e.g.*, to handle a soft boundary in a Merton model, or to regularize the step in a Heston model – which cannot be predicted by `NumGaussians` for the Monte Carlo engine. The input `more_randoms`, created using `Random_::Branch`, is provided for this contingency. Any such use will not be repeatable in bumped scenarios, so it must be recorded in the paths record.

We will ensure that the paths record endures longer than the workspace, so the workspace created by `NewWorkspace` is permitted to keep the input reference.

The `Envelope` method is used by the PDE to set up a grid; it gives upper and lower bounds to use for each state variable. `Envelope` is also called at the vol start time to find the initial state variables (though our convention is to use all zeroes).

The `StepAccumulator_` and `ModelStepper_` are constructed by the

SDE_ to its own specifications; they must communicate with each other, but each model can specialize that communication to its own ends. Similarly, if a **PathsRecord_** is used, it is constructed by and compatible with the accumulator.

10.9.1   *Valuation and Reevaluation*

At many stages of the valuation process, particularly in Monte Carlo path generation, we must be aware of the difference between base and bumped valuation. Data of various kinds must be captured during the former, and reused in the latter, case. We reflect this by creating a *re-evaluator* to support repeated valuation:

```
                        ─── Valuation.h ───
   class ReEvaluator_ : noncopyable
   {
   protected:
      Vector_<pair<String_, double> > baseVals_;
5  public:
      virtual ~ReEvaluator_();
      virtual Vector_<pair<String_, double> > Values
         (const SDE_* bumped_model = 0)
      const = 0;
10 };
```

A call to **Values()** returns the base values. Evaluation using the base model is performed inside the constructor of a derived class will compute the values, which will also store any information necessary to compute stable bumped values.[3]

In some advanced distributed applications, there is a slight performance gain from making ReEvaluator_ be Storable_, so that a base valuation on a single machine can be used to support bumped valuations on various remote machines. However, this is an onerous requirement which adds to the development effort for every new model.

10.10   Use Case Review: PDE

Let us see how these pieces fit together for a PDE pricing run. The flow of control proceeds as follows:

---

[3]Storing the base values as a `const public` member would seem more natural, but makes the derived class implementation awkward because they are naturally the last, not the first, quantity available to the constructor.

(1) The trade describes its underlyings.
(2) The model produces an SDE describing the dynamics of the necessary assets.
(3) The SDE produces a value request for use by the trade.
(4) The PDE solver wraps that value request in another, which will handle requests for past data and will provide stream locations to which values can be written.
(5) The trade produces a payout, filling out the value request in the process.
(6) The SDE produces an asset which will translate state variables to the requested observable index values.
(7) The PDE solver stores the SDE's steppers between event times (possibly interpolating event times of its own to limit the step size), and computes the envelope it will use.
(8) The PDE solver translates the payout's backward induction steps to actions manipulating the node values.
(9) The inner loop is run for each time step; see below.
(10) The rollback terminates at the vol start time, regardless of the reset time; at this point each slab is collapsed to a single value.
(11) Backward induction actions before the vol start time, if any, are applied to the vector of values, and payments from event times before the vol start time, if any, are added.
(12) The solver rescales the per-stream values to account for the discounting from vol start time to value date.

The inner loop is a dance of information exchange between abstract objects. At each spatial node:

(1) The solver sends the state variables to the asset.
(2) The asset updates index values and returns an update token.
(3) The solver sends this token, and a node values accessor, to the payout's.
(4) The payout uses index values to compute payment amounts, which it adds to the node values.
(5) The solver applies the backward induction actions.

If we have chosen to have the trade support PDE's more directly (see the discussion at the end of Sec. 10.3.3), then the payout can manipulate the node values rather than the solver.

Once the rollback is complete, we know the value of each stream and the payout provides the information for translating these to trade values.

PDE methods are generally stable, and the progress of a bumped run does not differ materially from that shown here; the main difference is that we will likely reuse the base envelope, especially if we are rescaling the grid during the run.

10.11   Use Case Review: Monte Carlo and Hedge

In a Monte Carlo, we must always distinguish between *base* and *bumped* valuations. The latter will reuse information from the base run in several ways, all aimed at preventing a discontinuous value change:

- To fix the order of eigenmodes, and other ordering issues.
- To fix AMC exercise decisions, which are otherwise chaotic.
- To repeat discrete jump or default events – we will reweight paths instead of generating different events.

The role of the paths record is to support the last of these needs, by making the history of the base valuation available during the bumped run. The Monte Carlo engine, which controls the AMC decider, is responsible for fixing decisions; this is accomplished by storing the base decisions in a fake decider which simply re-applies them. Eigenmode ordering and the like are handled by the SDE during creation of the stepper; for this purpose, the base stepper must be supplied so that a consistent bumped stepper can be created.

Given this, the flow of control proceeds as follows:

(1) The trade describes its underlyings.
(2) The model produces an SDE describing the dynamics of the necessary assets.
(3) The SDE produces a value request for use by the trade.
(4) The MC solver wraps that value request in another, which will handle requests for past data and will provide stream locations to which values can be written.
(5) The trade produces a payout, filling out the value request in the process.
(6) The SDE produces an asset which will translate state variables to index values, and will populate index paths as necessary.
(7) The MC solver stores the model's steppers between event times; for the bumped run, it supplies the stored stepper from the base run.
(8) The trade is evolved forward to the vol start time, and the accumulator is used to find the starting state variables, once only (not per path).

(9) For later event times, the inner loop is run for each path; see below.
(10) Backward induction actions are applied to the whole path set. These change values by creating new nodes within their streams.
(11) The node values are discounted, averaged across paths, and summed to produce stream values.
(12) The payout provides the weights to translate to trade values.

The inner loop is changed from the PDE case exactly as one would expect. The `operator=` member of `Payout_::State_` lets us initialize the state appropriately. For each event time on the path:

(1) The solver invokes the stepper to change the model state.
(2) The solver sends the state variables to the asset.
(3) The asset updates index values and index paths, and returns an update token.
(4) The solver invokes `DoDefault` as necessary, then `DoNode`, using this token.
(5) The payout uses index values and paths to compute payment amounts, which it adds to the node values.

10.11.1 *Causality*

The inner loop displayed here implements a *causal* Monte Carlo, in which the thread of execution moves forward in simulated financial time so that information from the future can never be used in a payout. Based on the same protocols, we can also write an *acausal* or *whole-path* Monte Carlo, where we take all the model steps, then circle back and compute the resulting payout at each event time. The only design constraint is that the `ValueRequest_` must not reuse addresses across different event times.

Acausal Monte Carlo, while less aesthetically appealing, may be faster in practice because it keeps a given piece of code – the stepper or the payout – in-process longer, reducing cache-swapping costs.

10.12   Costs and Benefits

This is the reward for the rigorous abstraction process we have followed in this chapter: the two methods are generic to the largest feasible class of models and of trades, and both access the model and trade in the same way. This is our path toward enabling the most powerful models: anything which creates a stepper and an asset is a model, and can be used to price any trade.

Many pet projects are casualties of this approach. There is no place here for tightly coupled methods which compute one particular hedge (generally an equity or FX delta) directly within the Monte Carlo; nor for highly model-specific methods such as PDEs for Asian options. Dependence on such highly specialized techniques is, in our opinion, a sign of a practitioner who has turned his back on the more important problem of computing a wide class of risks using a range of different models with maximum flexibility.

10.13 Assembling the Class Hierarchy

Here we work bottom-up, defining the most basic classes first. For a top-down view, where the motivation for a class is presented before its definition is known, it is advised to read this section backwards.

10.13.1 *Stepper*

Defined in Sec. 10.9.

10.13.2 *Asset Values and Tokens*

The update token defined in Sec. 10.6 is a *view* of underlying data which is manipulated by the model:

```
struct PathFixings_ : noncopyable
{
    Vector_<> vals_;
    Vector_<shared_ptr<IndexPath_> > paths_;
    Vector_<Handle_<IndexPath_> > pathRO_;
    int valMask_, dateMask_;

    Valuation::UpdateToken_ Token(const Time_& evt_t) const
    {
        return Valuation::UpdateToken_
                (vals_.begin(), pathRO_.begin(), valMask_,
                dateMask_, evt_t);
    }
};
```

This object is just a container of data; it does not know the time, or monitor how its contents are manipulated.

10.13.3 *SDE*

The SDE produces steppers, which advance the state in a Monte Carlo or supply coefficients to a PDE; and also provides the initial value request, and the assets which communicate index fixings to the trade.

```
――――――――――――――― SDE.h ―――――――――――
class SDE_ : noncopyable
{
public:
   virtual ~SDE_();
5  virtual ValueRequest_* NewRequest() const = 0;
   virtual Asset_* NewAsset(ValueRequest_& req) const = 0;

   virtual StepAccumulator_* NewAccumulator() const = 0;
   virtual ModelStepper_* NewStepper
10      (const Time_& from,
         const Time_& to,
         StepAccumulator_* cumulative,
         ModelStepper_* exemplar)
      const = 0;
15 };
```

The `StepAccumulator_` input to `NewStepper` is that returned from `NewAccumulator`; it can be `NULL` if the model needs no cumulative information across times or paths. The `exemplar` will be non-NULL only in bumped valuations, when it will contain the base-case output from `NewStepper`.

10.13.4 *Model*

The model's role is to produce an SDE for pricing, once the trade underlying is known. We choose to return this underlying as a reference, making the parent model responsible for its memory demands.

```
――――――――――――――― Model.h ―――――――――――
class Model_ : public Storable_
{
public:
   virtual Handle_<SDE_> ForTrade
5      (_ENV, const Underlying_& trade)
      const = 0;
   virtual Handle_<YieldCurve_> YieldCurve
```

```
          (const String_& ccy)
      const = 0;
10    virtual Model_* Mutant_Model
          (const String_& new_name,
           const Vector_<Handle_<Slide_> >& slides)
      const = 0;
};
```

We will discuss slides and `Mutant_Model` in Ch. 15.

10.13.5   *Trade*

The trade must state its underlyings, and also (given a value request) produce the payout. It turns out that trades are not `Storable_`; we will discuss this in Ch. 11, next.

10.13.6   *Historical Data Access*

The `ValueRequest_` provided by the model should not have to concern itself with requests for historical data; and the trade, in making requests, should not even be aware of whether the resulting fixings are historical or simulated. Thus we need to intercede between the two, supplying historical fixings to trades.

```
————————————————— ValueHistorical.h —————————————————
class PastAwareRequest_ : public ValueRequestImp_
{
    ValueRequestImp_& model_;
    const Environment_& env_;     // or other fixings access
5   Time_ resetTime_;
    long accountingDate_;

    // implement the construction of a single historical path
    Handle <IndexPathHistorical_> HistoricalPath
10      (const Index_& index,
         const Handle_<IndexPathHistorical_>& prior);

    struct History_ : IndexPathHistory_
    {
15      PastAwareRequest_* parent_;
        const IndexPathHistory_* base_;
        History_(PastAwareRequest_* p,
              const IndexPathHistory_* b)
        : parent_(p), base_(b) {}
```

```
        Time_ ResetTime() const
        {return parent_->resetTime_;}
        Handle_<IndexPathHistorical_> History
            (const Index_& index)
        const
        {
            auto prior = base_
                ? base_->History(index)
                : Handle_<IndexPathHistorical_>();
            return parent_->HistoricalPath(index, prior);
        }
    };

    Handle_<Payment::Tag_> PayDst
        (const Payment_& flow)
    {
        return flow.date_ < accountingDate_
            ? Payment::Null()
            : model_.PayDst(flow);
    }

    Handle_<Payment::Default::Tag_> DefaultDst
        (const String_& stream)
    {
        return model_.DefaultDst(stream);
    }

    address_t Fixing
        (const Time_& event,
         const Index_& index)
    {
        return event < resetTime_
            ? model_.InsertFixing(index.Fixing(0, event))
            : model_.Fixing(event, index);
    }
    address_t InsertFixing(double fixing)
    {
        return model_.InsertFixing(fixing);
    }

    IndexAddress_ IndexPath
        (const Time_& last_event_time,
         const Index_& index)
    {
```

```
65        History_ h(this, 0);
          return last_event_time < resetTime_
             ? model_.InsertPath
                 (HandleCast<IndexPath_>(h.History(index)))
             : model_.IndexPath(last_event_time, index, &h);
70      }
      IndexAddress_ IndexPath
          (const Time_& last_event,
           const Index_& index,
           const IndexPathHistory_* historical)
75      {
          History_ h(this, historical);
          return model_.IndexPath(last_event, index, &h);
      }
      IndexAddress_ InsertPath
80        (const Handle_<IndexPath_>& path)
      {
          return model_.InsertPath(path);
      }
};
```

It is not immediately obvious how `InsertFixing` and the three-argument form of `InsertPath` can ever be called for this class. However, it is good practice to have them simply forward to `model_`, so we will be able to nest `PastAwareRequest_` inside another decorator if need be.[4]

10.13.7 *Assets*

The asset is almost completely blank:

```
——————————————— Asset.h ———————————————
class Asset_ : noncopyable
{
public:
    virtual Valuation::UpdateToken_ Update
        (const Time_& event_time,
         const Vector_<>& state) = 0;
};
```

All our concrete assets will share an implementation scaffold, described in Sec. 13.3, which individual models will populate with model-specific local updaters.

---

[4]For example, this is used when forecasting expected payments, which is outside the scope of this volume.

10.13.8 *Solvers*

Each solver inherits from `ReEvaluator_`. The Monte Carlo solver exists in namespace `MonteCarlo`:

```cpp
/*─────────────── MC.cpp ───────────────*/
class Task_ : public ReEvaluator_
{
    scoped_ptr<ValueRequest_> request_;
    scoped_ptr<const Payout_> payout_;
    scoped_ptr<StepAccumulator_> cumulant_;
    Vector_<Handle_<ModelStepper_> > steps_;
    scoped_ptr<PathsRecord_> paths_;

public:
    Task_(const SDE_& model,
        ValueRequest_* request,
        const Payout_* payout);
};
```

The `request` and `payout` are *orphan* pointers, whose memory belongs to the `MC_` once the constructor is called.[5] We pass the `request` and `payout` separately, rather than have the task call `Trade_::Payout`, in order to avoid a compilation dependence of Monte Carlo tasks on `Trade_`s; the two should be separate and equal.

---

[5]Passing bare pointers may be considered bad form; in practice we will immediately catch them in `scoped_ptrs`.

Chapter 11

Standardized Trades

The fastest way to understand the protocols for numerical pricing is to look at a few trades, which of course will implement these protocols. We will return to non-numerical pricing in Ch. 14.

11.1   Trade Classes

We try to keep the trade's interface as narrow as possible, to preserve the maximum of implementation freedom.

```
                          ——————— Trade.h ———————
    class Trade_ : noncopyable
    {
    public:
        const Vector_<String_> valueNames_;
5       const Underlying_ underlying_;
        const String_ valueCcy_;

        Trade_(const Vector_<String_>& value_names,
               const Underlying_& underlying,
10             const String_& value_ccy);

        virtual Payout_* MakePayout
            (const Valuation::Parameters_& parameters,
             ValueRequest_& value_request)
15      const = 0;
    };
```

To avoid repeated evaluation of the underlying (and since there is no meaningful use of a trade that does not reference the underlying), it is stored in the base class upon construction. Note that there is no need for otiose accessor functions; const public data serves our purpose exactly and concisely. The valueNames_ are the keys of Payout_::StreamWeights from Sec. 10.8; Valuation::Parameters_ is a SETTINGS type – see Sec. 3.4 – containing discretization instructions which may be used by more complex trades.

The alert reader might wonder why a Trade_ is not Storable_. We do not store the trade itself, but another object containing the information from which it was made:

```
──────────────── Trade.h ────────────────
class TradeData_ : public Storable_
{
    mutable Handle_<Trade_> parsed_;
    virtual Trade_* XParse() const = 0;
 5 public:
    TradeData_(const String_& name) :
            Storable_("Trade", name) {}
    Handle_<Trade_> Parse() const;
    void Clear() const;    // un-Parse
10    Bookkeeping_ Bookkeeping() const;
};
```

This is the "trade" seen at the public interface by library users. The Parse function is called to convert it to an (immutable) Trade_. The Clear function may be called if necessary to free memory; parsed user-scripted trades are typically much larger in memory than their source scripts.

```
──────────────── Trade.cpp ────────────────
Handle_<Trade_> TradeData_::Parse() const
{
    // protection here if multithreading
    if (parsed_.Empty())
 5        parsed_.reset(XParse());
    return parsed_;
}
```

The Bookkeeping_ contains non-financial information about the trade, such as the trade date, counterparty, or nominal notional. These quantities do not enter the remainder of our discussion.

This interface is quite satisfactorily narrow: in fact, too narrow for many purposes. The missing functionality includes:

- More revealing representations than the rather opaque Payout_, for use in closed form pricing. We will address this with mixins; see Ch. 14.

- A query interface for introspection, *e.g.*, to identify range accrual trades.
- The `Storable_` interface itself, which would allow users to inspect the parsed `Trade_` object directly. See Sec. 5.6 for a partial fix.

To support storage of trades, we create a mark-up file (see Sec. 5.4).

```
——————————— Trade.abstract.storable.if ———————————
BUILDS Handle_<TradeData_>
```

11.2 Cash

The simplest possible derivative trade is an agreement to pay deterministic cash flows at some set of future dates. We need to implement this only for flows in a single currency; multi-currency flows can be formed as composite trades (see Sec. 11.8) or, more likely, booked as separate transactions.

Thus the trade's data during pricing are quite simple:

```
——————————— CashTrade.cpp ———————————
struct CashTrade_ · public Trade_
{
    Vector_<Flow_> flows_;
```

We use `struct` rather than `class`, since we will keep the entire implementation inside a source file and publish only a factory function in the header. The other necessary data is stored directly in `Trade_`.

```
——————————— CashTrade.cpp ———————————
     CashTrade_(const String_& name,
             const String_& ccy,
             const Vector_<Flow_>& flows)
         :
5        Trade_(Vec1(name), CashUnderlying(ccy, flows), ccy),
         flows_(flows)
         {   }

         Payout_* MakePayout
10           (const Valuation::Parameters_&,
             ValueRequest_& value_request)
         const;
};
```

The function `CashUnderlying` constructs an `Underlying_` with the appropriate currency and last payment date, extracted from the `flows`.

Now we can write the `Payout_` (see Sec. 10.8) which will be used in numerical pricing. (Using a numerical method for this purpose is obviously overkill – but ensuring its efficiency is a test of the quality of our framework.)

```cpp
                              CashTrade.cpp
struct CashPayout_ : Payout_    // clumsy implementation
{
    String_ name_;    // of trade and of stream
    vector<pair<dst_t, double> > flows_;

    CashPayout_(const String_& name) : name_(name) {}

    Vector_<Time_> EventTimes() const
    { return Vec1(Time::Minimum()); }

    void DoNode
        (const Valuation::UpdateToken_&,
         State_*,
         NodeValues_& pay_dst)
    const
    {
        for (auto pf = flows_.begin();
                   pf != flows_.end(); ++pf)
        {
            pay_dst[pf->first] += pf->second;
        }
    }

    Vector_<BackwardInduction::Action_> BackwardSteps()
        const
    { return Payout_::BackwardSteps(); }

    map<String_, Vector_<pair<String_, double> > >
        StreamWeights() const
    {
        map<String_, Vector_<pair<String_, double> > > r;
        r[name_] = Vec1(make_pair(name_, 1.0));
    }
};
```

The cashflows of the trade become `pair`s of tag and amount, since the tag encapsulates all information except the amount about a trade's payment. The payments will be made into a stream whose name is the same as the trade's; thus `StreamWeights` needs only to identify that fact. We must create a nominal event time, which we set in the dim past. We expect

DoNode to be called for that event time (and only that one), at which point we make all our payments.

The last two functions in the above class promise restrictions – no backward induction, and a single stream for the trade – that are common to most simple trades. So it is best to put them in an implementation base class and reuse their code.

```
————————————— PayoutEuropean.h —————————————
class EuropeanPayout_ : public Payout_
{
protected:
    const String_ name_;    // of trade and of stream
5   EuropeanPayout_(const String_& name) : name_(name) {}

    Vector_<BackwardInduction::Action_> BackwardSteps()
        const
    { return Vector_<BackwardInduction::Action_>(); }
10
    map<String_, Vector_<pair<String_, double> > >
        StreamWeights() const
    {
        map<String_, Vector_<pair<String_, double> > > r;
15      r[name_] = Vec1(make_pair(name_, 1.0));
    }
};
```

Now **CashPayout_** can derive from this.

The value request is the tool that lets the trade create a payout, by providing it with promised addresses and payment tags. But we have to convert the **Flow_** objects held by the trade into **Payment_** objects which communicate the context of the cashflow.

```
————————————— CashTrade.cpp —————————————
Payment  MakePayment
    (const String_& stream,
    const String_& ccy,
    const Flow_& flow)
5 {
    static const Time_ WHEN = Time::Minimum();
    return Payment_(WHEN, ccy, flow.payDate_, stream,
        Payment::Info_("Contractual cashflow", WHEN));
}
```

We could make this a member, rather than a nonmember to which class data are passed; but we habitually prefer to keep classes small and function interfaces explicit, even at some cost in verbosity.

```
──────────── CashTrade.cpp ────────────
Payout_* CashTrade_::MakePayout
   (const Valuation::Parameters_&,
    ValueRequest_& value_request)
const
{
    const String_& name = valueNames_[0];
    auto_ptr<CashPayout_> retval(new CashPayout_(name));
    for (auto pf = flows_.begin();
         pf != flows_.end(); ++pf)
    {
        Payout_::dst_t tag = value_request.PayDst
               (MakePayment(name, valueCcy_, *pf));
        if (tag != Payment::Null())
            retval->flows_.push_back
                   (make_pair(tag, pf->amount_));
    }
    return retval.release();
}
```

This shows the use the value request, and of **Payment::Null** to eliminate
worthless cashflows – the latter is an optimization[1] only, and not necessary
for correct pricing or reporting.

```
──────────── CashTrade.1.storable.if ────────────
ISA Trade
BUILDS Handle_<TradeData_>
?STRING name
STRING ccy
NUMBER[] amounts
' Amounts we receive, negative if we pay
INTEGER[] dates
' Dates on which we receive a corresponding amount
CONDITION {data->amounts.size() == data->dates.size()} \
    {flow dates and amounts must have the same size}
```

This mark-up supports the trade data, which is extremely similar to the
CashTrade_ used for pricing:

```
──────────── CashTrade.cpp ────────────
struct CashTradeData_ : TradeData_
{
    String_ ccy_;
    Vector_<Flow_> flows_;
    CashTradeData_(const String_& name,
```

───────────────────

[1]Not a very crucial one.

```
            const String_& ccy,
            const vector<Flow_>& flows)
        :
      TradeData_(name),
10    ccy_(ccy),
      flows_(flows)
      {   }
      Trade_* XParse() const
          {return new CashTrade_(name_, ccy_, flows_);}
15  };
```

The duplication of members in `CashTrade_` and `CashTradeData_` is the price we pay for control over `Parse` and `Clear`.

11.3   Equity and FX

A slightly more interesting trade is an equity forward – a contract to receive an equity index (or the cash value thereof) in exchange for a fixed amount at a forward date. Now that we have seen the relationship of `Payout_`, `Trade_`, and `TradeData_`, we can write them from the inside out – this is how we will usually proceed when creating a new trade.

11.3.1   *Equity Forward Payout*

What must the payout look like?

```
                  ———— EquityTrade.cpp ————
struct EquityForwardPayout_  : EuropeanPayout_
{
    Time_ expiry_;
    Valuation::address_t fixing_;
5   double strike_, size_;    // size is signed
    dst_t dst_;

    EquityForwardPayout_
        (const String_& name, const Time_& expiry,
10       Valuation::address_t fixing, double strike,
         double size, const Handle_<Payment::Tag_>& dst);

    Vector_<Time_> EventTimes() const
    { return Vec1(expiry_); }
15
    void DoNode
        (const Valuation::UpdateToken_& values,
```

```
            State_*,
            NodeValues_& pay)
20      const
        {
            assert(values.eventTime_ == expiry_);
            const double spot = values[fixing_];
            pay[dst_] += size_ * (spot - strike_);
25      }
    };
```

This illustrates the extraction of the necessary fixing – the index spot value at expiry – from the `values`.[2] The trade must now have the necessary data to construct this payout:

```
————————————— EquityTrade.cpp ———————
Payout_* EquityForward_::MakePayout
    (const Valuation::Parameters_&,
    ValueRequest_& mkt)
const
5 {
    const String_& name = valueNames_[0];
    Handle_<Payment::Tag_> payDst
            (mkt.PayDst
                (MakePayment
10                   (expiry_, valueCcy_, name)));
    return new EquityForwardPayout_
        (name, expiry_, mkt.Fixing(expiry_, *eqIndex_),
        strike_, size_, payDst);
}
```

Here we have encapsulated two tasks – forming the index, and describing the payment – in the member `eqIndex_` and the function `MakePayment` respectively. The latter simply marshals other member data:

```
————————————— EquityTrade.cpp ———————
Payment_ MakePayment
    (const Time_& expiry,
    const String_& ccy,
    const String_& stream)
5 {
    Payment::Info_ tag("Equity forward delivery", expiry);
    return Payment_
```

---

[2]Our generation of a single cash payment is equivalent to assuming cash settlement. A cash system might have to make a finer distinction, by treating the stock delivery the same as a cashflow.

```
                     (expiry, ccy, expiry.Julian(), stream, tag);
}
```

11.3.2 *Equity Index*

A spot equity price is a simple piece of market data, so of course we will
have a class derived from `Index_` to represent it. It is best to allow for
observation of equity forward as well as spot prices, though we do not need
them for this trade.

```
─────────────────── IndexEquity.h ───────────────────
namespace Index
{
   class Equity_ : public Index_
   {
      Cell_ delivery_;    // empty, date, or increment
      String_ Name() const;
   public:
      const String_ eqName_;
      long ExpiryDate(long fixing_date) const;

      // can't supply both date and increment!
      Equity_(const String_& eq_name,
            const long* delivery_date = 0,
            const String_* delay_increment = 0);
   };
}
```

If we are using the index-naming conventions of Sec. 9.1, we must produce a name following those conventions:

```
─────────────────── IndexEquity.cpp ───────────────────
String_ Index::Equity_::Name() const
{
   String_ ret = "EQ[" + eqName_ + "]";
   if (Cell::IsString(delivery_))
      ret += ">" + Cell::AsString(delivery_);
   else if (Cell::IsNumber(delivery_))
      ret += "@" + String::FromDate
            (AsInt(Cell::AsNumber(delivery_)));
   return ret;
}
```

11.3.3 *Equity Forward Data*

The equity forward trade thus takes an equity name and constructs the index it will use in pricing:

```
─────────────── EquityTrade.cpp ───────────────
EquityForward_::EquityForward_
    (const String_& trade_name,
    const String_& eq_name,
    const String_& ccy,
5   const Time_& expiry,
    double strike,
    int signed_num_contracts)
    :
Trade_(Vec1(trade_name), EquityUnderlying(eq_name), ccy),
10  eqIndex_(new Index::Equity_(eq_name)),
expiry_(expiry),
strike_(strike),
size_(signed_num_contracts)
{   }
```

11.3.4 *FX Option*

The pricing of simple FX trades is conceptually similar to that of the corresponding equity trades, and the FX forward very closely resembles the equity forward just described. We can distinguish between cash and physical settlement:

```
─────────────── FxTrade.cpp ───────────────
// cash settlement
void FxForwardPayout_::DoNode
    (const Valuation::UpdateToken_& values,
    State_*,
5   NodeValues_& pay)
const
{
    assert(values.eventTime_ == expiry_);
    const double spot = values[fixing_];
10  pay[dst_] += domAmt_ + spot * fgnAmt_;
}
```

```
─────────────── FxTrade.cpp ───────────────
// physical settlement -- preferred
void FxForwardPayout_::DoNode
    (const Valuation::UpdateToken_& values,
    State_*,
5   NodeValues_& pay)
```

```
const
{
   assert(values.eventTime_ == expiry_);
   pay[domDst_] += domAmt_;
   pay[fgnDst_] += fgnAmt_;
}
```

The latter approach will generate a clearer payment report, and more accurately captures the nature of an FX forward. Thus, when creating FX options, we will follow the same path. In the pricing implementation, we use signed amounts rather than enumerated flags:

```
——————————— FxTrade.cpp ———————————
struct FxOptionPayout_: EuropeanPayout_
{
   Time_ expiry_;
   Valuation::address_t fixing_;
   double domAmt_, fgnAmt_;
   dst_t domDst_, fgnDst_;
   int sign_;    // long-short flag

   FxOptionPayout_
       (const String_& name, const Time_& expiry,
        Valuation::address_t fixing, double dom_amt,
        dst_t dom_dst, double fgn_amt, dst_t fgn_dst,
        int sign);

   Vector_<Time_> EventTimes() const
   { return Vec1(expiry_); }

   void DoNode
       (const Valuation::UpdateToken & values,
        State_*,
        NodeValues_& pay)
   const
   {
       assert(values.eventTime_ == expiry_);
       const double spot = values[fixing_];
       if (sign_ * (domAmt_ + spot * fgnAmt_) > 0.0)
       {   // in the money
           pay[domDst_] += domAmt_;
           pay[fgnDst_] += fgnAmt_;
       }
   }
};
```

At the public interface, however, we favor positive notional amounts and flags.

```
——————————— FxOption.1.storable.if ———————————
 ISA Trade
 BUILDS Handle_<TradeData_>
 ?STRING name
 STRING dom_ccy
5 NUMBER dom_amt
 ' Domestic notional, positive
 STRING fgn_ccy
 NUMBER fgn_amt
 ' Foreign notional, positive
10 RecPay rec_pay_fgn
 ' Determines whether foreign notional is received or paid
 TIME expiry
 ' Date on which FX spot is observed and exercise decided
 ?INTEGER delivery
15 ' Date on which FX is paid
 CONDITION {dom_amt > 0.0 && fgn_amt > 0.0}
     {Notional amounts must be positive}
```

The `delivery` date is optional, because in most cases it can be deduced from the exercise date.

11.3.5 *Forcing Backward Induction*

We can write the payout another way, exercising the optionality during a backward induction sweep rather than immediately. This provides no implementation advantage, but illustrates the use of the backward induction protocols in a simple case.

```
——————————— FxTrade.cpp ———————————
 struct FxOptionPayout_AMC_ : Payout_
 {
    FxForwardPayout_ underlying_;
    Time_ expiry_;
5   int sign_;
    amount_t spotDst_;   // observable for AMC

    Vector_<Time_> EventTimes() const
    {   return Vec1(expiry_);   }
10
    State_* NewState() const {return 0;}
    void StartPath(State_*) const {}
```

```
     void DoNode
15      (const Valuation::UpdateToken_& values,
         State_* state,
         NodeValues_& pay_dst)
     const
     {
20      underlying_.DoNode(values, state, pay_dst);
        pay_dst[spotDst_] = values[underlying_.fixing_];
     }

     Vector_<BackwardInduction::Action_> BackwardSteps()
25   const
     {
        BackwardInduction::Action_ retval
              (underlying_.Stream(), expiry_,
               expiry_.Julian());
30      BackwardInduction::Exercise_ bermEx;
        bermEx.sign_ = sign_;
        // observe the spot -- see DoNode
        bermEx.observables_.push_back(spotDst_);
        retval.details_ = bermEx;
35      return Vec1(retval);
     }
};
```

The backward-induction action will terminate the stream when it gains value by doing so (if `sign_` is positive); its decision is based on a single observable, the spot FX price, which is all we need. There are no fees and no underlying streams received on exercise, because we are representing the trade as an FX forward with an option to terminate. The stream shares the trade's name and receives all the value, as expected by `EuropeanPayout_`. In `MakePayment`, we will set the payment's `tag_.conditions_.exerciseCondition_` to `ON_CONTINUATION`, flagging that it is no longer an unconditional cashflow.

If we chose to implement the payment as a fee, we would remove the payment from the value stream and instead append the payment tag to `bermEx->fees_`, and set the payment condition to `ON_EXERCISE`.

11.4   Legs and Swaps

We begin with the description of a fixed leg trade, alluded to in Sec. 8.5.2. The trade data is:

```
────────────────── FixedLegTrade.1.storable.if ──────────────────
ISA TradeData
BUILDS Handle_<TradeData_>
?STRING name
STRING ccy
5│ LegScheduleParams terms
STRING rec_pay
NUMBER[] notional
NUMBER[] coupon_rate
?STRING notional_exchange
```

To best support the creation of numerical payouts for this and other fixed-leg trades, we extend `EuropeanPayout_` to support the accumulation of leg-based flows. First we need a generic way to represent a (possibly) stochastic amount:

```
──────────────────────── TradeAmount.h ────────────────────────
class TradeAmount_ : noncopyable
{
public:
    virtual ~TradeAmount_();
5│   typedef Valuation::UpdateToken_ UpdateToken_;
    virtual double operator()
        (const UpdateToken_& values) const = 0;
};
```

The most common amount is simply a wrapper around a single fixing:

```
────────────────────── TradeAmountUtils.h ──────────────────────
namespace TradeAmount
{
    using Valuation::UpdateToken_;
    struct Fixing_ : TradeAmount_
5│   {
        const Valuation::address_t loc_;
        Fixing_(Valuation::address_t p) : loc_(p) {}
        double operator()(const UpdateToken_& values) const
        { return values[loc_]; }
10│  };
```

More complex amounts will be needed, *e.g.*, for payments incorporating a margin or spread. At a minimal efficiency cost, we can combine constant and spread amounts:

```
———————————————————— TradeAmountUtils.h ——————————
    struct Spread_ : TradeAmount_
    {
        Handle_<TradeAmount_> base_;
        const double spread_;
        Spread_(double s) : spread_(s) {}    // no base
        Spread_(double s, const Handle_<TradeAmount_>& b);
        double operator()(const UpdateToken_& values) const
        {
            return base_.Empty()
                   ? spread_
                   : spread_ + (*base_)(values);
        }
    };
}
```

These are all the ingredients we need to start creating payouts. We define
a coupon payment, then a `Payout_` composed of coupons:

```
———————————————————————— LegBased.h ——————————————
namespace LegBased
{
    struct Coupon_
    {
        Time_ eventTime_;
        Handle_<TradeAmount_> rate_;
        double dcf_;
        Handle_<Payment::Tag_> pay_;
    };

    class Payout_ : public EuropeanPayout_
    {
        // coupons indexed by own event time
        multimap<Time_, Coupon_> coupons_;
    public:
        Payout_(const String_& trade_name)
            : EuropeanPayout_(trade_name) {}

        operator+=(const Coupon_& c);

        Vector_<Time_> EventTimes() const
        { return Unique(Keys(coupons_) );}

        void DoNode
            (const Valuation::UpdateToken_& vls,
```

```
             State_* state,
             NodeValues_&dst)
        const
        {
30          auto tt = coupons_.equal_range(vls.eventTime_);
            for (auto pc = tt.first; pc != tt.second; ++pc)
            {
                const Coupon_& c = pc->second;
                dst[c.pay_] += c.dcf_ * (*c.rate_)(vls);
35          }
        }
    };
}
```

Now forming a numerical payout, for any leg-based trade, is just a matter of creating the `Coupon_` amounts. We simplify this task with a support function which converts a `LegPeriod_` (Sec. 8.5) to a `LegBased::Coupon_`. In order that this function be extensible to unforeseen `CouponRate_` types, we supply two helper classes in **namespace LegBased**:

LegBased.h

```
struct MakeRate_ : noncopyable
{
    virtual pair<Time_, Handle_<TradeAmount_> > operator()
        (ValueRequest_& request,
5         const CouponRate_& rate)
    const;
};

struct MakeCoupon_ : unary_function<LegPeriod_, Coupon_>
10 {
    ValueRequest_& valueRequest_;
    const Handle_<MakeRate_> makeRate_;
    const String_ tradeName_;
    const String_ payCcy_;

15
    MakeCoupon_
        (ValueRequest_& v, const Handle_<MakeRate_>& r,
         const String_&tn, const String_& ccy);
    virtual ~MakeCoupon_();

20
    virtual Coupon_ operator()
        (const LegPeriod_& period)
    const;
};
```

The default implementation `MakeRate_::operator()` recognizes the rate types – fixed, Libor, Libor plus margin, and interpolated stub Libor – discussed in Secs. 8.5 and 8.5.1. Other trades can reuse `MakeCoupon_` by defining their own classes, derived from `MakeRate_`, which will handle the trade's own rate type. Most rate types can be handled this way, without the need to override the implementation of `MakeCoupon_::operator()`.

So that basic rate functionality can be accessed without explicitly creating a `MakeRate_` object – which is stateless anyway – we provide the utility operator

```
 ──────── LegBased.cpp ────────
const LegBased::MakeRate_& operator+
    (const Handle_<LegBased::MakeRate_>& mr)
{
    static const LegBased::MakeRate_ DEFVAL;
5   return mr.Empty() ? DEFVAL : *mr;
}
```

This simplifies the implementation of `MakeCoupon_` as well as of derived `MakeRate_` classes.

```
 ──────── LegBased.cpp ────────
LegBased::Coupon_ LegBased::MakeCoupon_::operator()
    (const LegPeriod_& period)
const
{
5   Coupon_ ret;
    ret.dcf_ = period.accrual_->dcf_;
    Payment_ pay;
    pay.tag_.period_.Set(*period.accrual_);
    pay.tag_.description_ = "Coupon payment";
10  pay.ccy_ = payCcy_;
    pay.date_ = period.payDate_;
    pay.stream_ = TradeName_;
    tie(pay.eventTime_, ret.rate_)
        = (+makeRate_)(valueRequest_, *period.rate_);
15  ret.eventTime_ = pay.tag_.knownTime_ = pay.eventTime_;
    ret.pay_ = valueRequest_.PayDst(pay);
    return ret;
}
```

11.4.1 *Putting it Together*

We have found that a fixed leg trade contains just a leg, and value currency:

```
───────────────────── LegTrade.cpp ─────────────
class FixedLeg_ : public Trade_
{
   Vector_<LegPeriod_> leg_;
   String_ ccy_;
public:
   Payout_* MakePayout
      (const Valuation::Parameters_&,
       ValueRequest_& value_request)
   const
   {
      static const Handle_<LegBased::MakeRate_> DMR;
      auto_ptr<LegBased::Payout_> retval
            (new LegBased::Payout_(valueNames_[0]));
      LegBased::MakeCoupon_ makeCoupon
            (value_request, DMR, valueNames_[0], ccy_);
      for (auto p = leg_.begin(); p != leg_.end(); ++p)
         *retval += makeCoupon(*p);
      return retval.release();
   }
};
```

Precisely the same approach works for Libor legs, with or without stubs and margins. The trade is just a wrapper which mediates between the leg-construction functions of Sec. 8.5 and the **LegBased** payout.

11.5 Caps

To construct a cap, we first need to introduce the rate it pays[3]:

```
───────────────────── Cap.h ─────────────
struct CapRate_ : CouponRate_
{
   Handle_<CouponRate_> underlying_;
   double strike_;
   CapFloor_ type_;
   BuySell_ sign_;
};
```

Now we provide a derived class of **MakeRate_** which will handle this. We would like to handle caps on anything, not just on Libor, which complicates the implementation slightly.

---

[3]The type **CapFloor_** is another machine-generated enumeration. We must be careful not to confuse caps with calls generally; a cap is a call on rates but a put on bonds.

```
────────────────── Cap.h ──────────────────
   struct MakeRateCapped_ : LegBased::MakeRate_
   {
      Handle_<MakeRate_> base_;
      MakeRateCapped_() {}
5     MakeRateCapped_(const Handle_<MakeRate_>& base)
         : base_(base) {}

      pair<Time_, Handle_<TradeAmount_> > operator()
         (ValueRequest_& vls, const CouponRate_& lr) const;
10 };
```

```
────────────────── Cap.cpp ──────────────────
   pair<Time_, Handle_<TradeAmount_> >
   MakeRateCapped_::operator()
      (ValueRequest_& model, const CouponRate_& rate)
   const
5  {
      if (DYN_PTR(cap, const CapRate_, &rate))
      {
         pair<Time_, Handle_<TradeAmount_> > temp
               = (+base )(model, *cap->underlying );
10       Handle_<TradeAmount_> capAmt
               (NewCapAmount(temp.second, cap->strike_,
                     cap->type_, cap->sign_.Sign()));
         return make_pair(temp.first, capAmt);
      }
15    return (+base_)(model, rate);
   }
```

A default-constructed `MakeRateCapped_` will support caps on Libor; to cap a more exotic rate, we need only provide the engine (derived from `MakeRate_`) which computes fixings for that rate.

The construction of a Libor cap trade should now be entirely predictable. We begin with the trade data:

```
────────────────── LiborCapTrade.1.storable.it ──────────────────
   ISA TradeData
   BUILDS Handle_<TradeData_>
   ?STRING name
   STRING ccy
5  LegScheduleParams terms
   ?STRING fixing_hols
   ' If omitted, use currency defaults
   ?NUMBER[] fixing_delay
   STRING buy_sell
```

```
10  STRING cap_floor
    NUMBER[] notional
    NUMBER[] strike
```

This is sufficient data to call a function **LegBuild::LiborCap**, which will in turn call **LegBuild::Libor** (see Sec. 8.5) and then build a new leg with **CapRate_**s replacing the Libor rates.

11.6   Swaps and Swaptions

There is not much more to say about swaps: they are collections of one or more (but almost always two) legs, for each of which we will store data, build a leg, and add its payments to **LegBased::Payout_** for pricing. We may choose to implement swaps as a subclass of composite trade; see Sec. 11.8.

A first cut of a European swaption would build the underlying swap as another **LegBased::Payout_**, then implement the option on this underlying during a backward induction phase (in Monte Carlo) or postprocessing step (in PDE); this would resemble the FX option payout in Sec. 11.3.5. However, this involves simulating the model state at each Libor fixing date of the underlying swap; thus it will be far faster, and also more precise, to create a payout which forecasts the Libor rates at the swaption's expiry.

If the underlying swap is known to be standard, so that its rate and sensitivity can be described as indices (as in Sec. 9.1), then we can assign their computation to the model and write a very simple payout. This also allows optimizations on the model side (see Sec. 13.3.3 for an example), and is necessary for cash-settled swaptions, though we do not show their implementation here.

In general, though, we must consider options on swaps with time-varying coupons or notionals, which cannot be passed off to the model. We need to use the methods of Sec. 11.4, but creating *forecasts* of Libor at the swaption expiry rather than waiting for *fixings*.

Fortunately, we can accomplish this with an overload of **LegBased::MakeRate_**:

```
──────────── Swaption.cpp ────────────
    struct MakeForecast_ : LegBased::MakeRate_
    {
        Time_ expiry_;
        MakeForecast_(const Time_& expiry) : expiry_(expiry) {}
5       pair<Time_, Handle_<TradeAmount_> > operator()
            (ValueRequest_& req, const CouponRate_& rate) const
```

```
         {
             if (DYN_PTR(libor, const LiborRate_, &rate))
             {
10               IndexIr_ index(libor->ccy_, libor->tenor_);
                 return make_pair(expiry_,
                     AsAmount(req.Fixing(expiry_, index)));
             }
             return LegBased::MakeRate_::operator()(req, rate);
15       }
     };
```

Now calling `LegBased::MakeCoupon_` with a `MakeForecast_` will convert the legs of the underlying swap into a series of `TradeAmount_s`. This is still not good enough for our purposes: rather than pay the various coupons into a stream, we must obtain and sum their discounted values. Discounting on the trade side is somewhat unusual, but does allow more efficient support of some kinds of optionality.

Thus we replace `MakeCoupon_` altogether, with a function which will create discounted values for us to accumulate:

```
                  ———— Swaption.cpp ————
struct ToDiscountedValue_
        : unary_function<LegPeriod_, TradeAmount_*>
{
    ValueRequest_& valueRequest_;
5   const String_ tradeName_;
    const String_ payCcy_;
    const MakeForecast_ makeRate_;

    TradeAmount_* operator()(const LegPeriod_& pd) const
10  {
        Handle_<TradeAmount_> rate = makeRate_
                (valueRequest_, *pd.rate_).second;
        IndexDf_ df(payCcy_, pd.payDate_);
        Valuation::address_t dfLoc = valueRequest_.Fixing
15          (makeRate_.expiry_, df);
        return new DiscountedValue_
                (rate, dfLoc, pd.accrual_->dcf_);
    }
};
```

Here `DiscountedValue_` is a simple local structure which evaluates the `rate`, looks up the discount factor from the update token (provided to `TradeAmount_::operator()`) using `dfLoc`, and multiplies them together with the daycount fraction to produce its value.

A numerical swaption trade contains a list of such values, which represent the underlying:

```Swaption.cpp
struct SwaptionPayout_ : EuropeanPayout_
{
    Time_ expiry_;
    int sign_;    // buy/sell
5   Vector_<Handle_<TradeAmount_> > underlyings_;
    dst_t payDst_;

    SwaptionPayout_(const Time_& expiry, int sign,
        const Vector_<Handle_<TradeAmount_> >& underlyings,
10       const dst_t& pay_dst);
    Vector_<Time_> EventTimes() const
    { return Vec1(expiry_); }

    void DoNode
15      (const Valuation::UpdateToken_& values,
         State_*,
         NodeValues_& pay)
    {
        assert(values.eventTime_ == expiry_);
20       double pv = 0.0;
        for (auto pu = underlyings_.begin();
                  pu != underlyings_.end(); ++pu)
            pv += (**pu)(values);
        pay[payDst_] += sign_ * Max(0.0, pv);
25   }
};
```

11.7   Bermudans

Pricing European swaptions in this way almost prepares us for Bermudan options. The only remaining step is to set up the backward induction actions which will support American Monte Carlo pricing. We must decide what observables to supply – including an indicator of the state of volatility, if we are using a stochastic-vol model – and form the backward induction actions as described in Sec. 10.7.

11.7.1 *Two Views*

This procedure gives us a Bermudan option, viewed as a union of European swaptions. Another, equally valid approach is to price a callable swap directly: we use the methods of Sec. 11.4 to define the swap payments, and then insert backward induction actions which, upon exercise, terminate the payment stream.

There is no strong reason to prefer either of these approaches over another. The union-of-swaptions method, since it forecasts exercise values at each exercise date, reduces numerical noise at the cost of a increased computation per node. We should be aware of both approaches, so that we can price more complex callable swaps directly without the need to artificially divide them into swap and option components.

11.8 Composites

So far we have considered trades as independent entities, each producing a single value. There are several reasons we might need a more general view:

- A trade which makes identical payments, but on varying notional amounts, to many counterparties (*e.g.*, retail investors);
- A sum of trades, to be considered as a single trade for some purposes (*e.g.*, counterparty analysis);
- A collection of trades, to be simultaneously valued in a single Monte Carlo run for optimization.

11.8.1 *Rescaled Trades*

The first case, where only the `StreamWeights` are changed, is simple to implement. The data model for the trade is obvious:

RescaledTrade.1.storable.if

```
   ISA Trade
   BUILDS Handle_<TradeData_>
   ?STRING name
   Trade security
5    The underlying trade (presumed to have unit notional)
   STRING[] client_names
   NUMBER[] client_notionals
   CONDITION {data->client_names.size() == \
      data->client_notionals.size()} \
10    {client names and notionals must have the same size}
```

```
CONDITION {Unique(data->client_names).size() == \
   data->client_names.size()} \
   {client names must all be distinct}
```

The rescaled `Payout_` simply forwards everything to the `security`'s `Payout_`, and overrides only `StreamWeights`. We begin by creating a base class for decoration of `Payout_` (see Sec. 2.6):

```
——————————— DecoratePayout.h ———————————
   struct DecoratedPayout_ : Payout_
   {
      Handle_<Payout_> base_;
      DecoratedPayout_(const Handle_<Payout_>& base)
5     : base_(base) {}

      Vector_<Time_> EventTimes() const
      { return base_->EventTimes(); }
      State_* NewState() const {return base_->NewState();}
10    // ...
   };
```

This both simplifies a concrete decorator implementation, and reduces the chance of failure-to-forward errors due to impure virtual functions in `Payout_`. The payout is now quite simple:

```
——————————— RescaledTrade.cpp ———————————
   struct RescaledPayout_ : DecoratedPayout_
   {
      map<String_, double> scalers_;
      RescaledPayout_(const Handle_<Payout_>& base,
5           const map<String_, double>& scalers)
       : DecoratedPayout_(base), scalers_(scalers) {}

      map<String_, Vector_<pair<String_, double> > >
         StreamWeights() const
10    {
         map<String_, Vector_<pair<String_, double> > > ret,
            base = base_->StreamWeights();
         REQUIRE0(base.size() == 1,
               "Trade already has multiple payments");
15       for (auto pc = scalers_.begin();
               pc != scalers_.end(); ++pc)
         {
            RescaleSecond_<String_, double> s(pc->second);
            ret[pc->first]= Apply(s, base.begin()->second);
20       }
```

```
        return ret;
    }
};
```

The concept of paying the same value to multiple counterparties only makes sense if there is a single "same value": thus we insist that the base trade must generate only one value. A related concept, rescaling a trade with (potentially) many payouts by a single constant multiplier, is easily implemented but does not seem to be necessary in practice.

Once the trade data (the `.if` file above) and payout are defined, creation of the `Trade_` and `TradeData_` are straightforward.

11.8.2 *Sums and Collections*

True composites, containing several trades, require a little more implementation effort. We can share a data model, distinguishing *collection trades* from *sum trades* based on whether the final values are distinct, or are summed to produce a single value.

```
──────────────── CompositeTrade.1.storable.if ────────────────
ISA Trade
BUILDS Handle_<TradeData_>
?STRING name
+Trade components
5    The trades to collect or sum
BOOLEAN sum_payouts
      If false, payouts are distinct and reported separately
```

The `TradeData_` and `Trade_` are now very easy to implement, but the `Payout_` has some subtleties.

First, what if two component trades make payments into the same stream – or, more accurately, if the names of two streams coincide?[4] We cannot let the streams mix, which would confuse the trade values and most likely lead to double-counting. Thus during creation of the member payouts, we intervene in the requests made by member trades:

```
──────────────── CompositeTrade.cpp ────────────────
struct XValuesForComposite_ : ValueRequest_
{
    ValueRequest_& base_;
    String_ prefix_;
5
```

---

[4]We have so far used only streams with the same name as the trade, but this is not a universal law.

```
     XValuesForComposite_(ValueRequest_& vr) : base_(vr) {}

     Handle_<Payment::Tag_> PayDst
        (const Payment_& flow)
10   {
        Payment_ temp(flow);
        temp.stream_ = prefix_ + temp.stream_;
        return base_.PayDst(temp);
     }

15
     Handle_<Payment::Default::Tag_> DefaultDst
        (const String_& stream)
     {
        return base_.DefaultDst(prefix_ + stream);
20   }
     // ...
};
```

The stream prefixes are decided by the `Trade_` during creation of the `CompositePayout_`, so we can also store them in the payout. We will likely use a simple system of prefixes such as `"[0]"`, `"[1]"`, etc.[5]

The collection's state must store substates for every member payout. We simplify this with a template `Composite_` class (see Sec. 16.4).

CompositeTrade.cpp

```
struct CompositePayout_ : Payout_
{
    Vector_<Handle_<Payout_> > vals_;
    Vector_<Vector_<Time_> > eventTimes_;
5   Vector_<String_> streamPrefixes_;
    Maybe_<String_> sumName_;    // empty for collections
    const Vector_<Handle_<Payout_> >& Trades() const
    { return vals_; }

10  // handle state with a Composite_
    typedef Composite_<State_> state_t;
    State_* NewState() const
    {
        auto_ptr<state_t> retval(new state_t);
15      for (auto pt = Trades().begin();
                pt != Trades().end(); ++pt)
        {
```

---

[5]In some advanced applications we might have to wrap a composite around a pre-existing payout, which might itself be composite; to support this, we must put more unique information into the prefixes so that we can safely handle their occasional absence.

```
                retval->Append((*pt)->NewState());
            }
20          return retval.release();
        };

        void StartPath(State_* _state) const
        {
25          state_t& state = CoerceComposite(_state);
            assert(state.Size() == Trades().size());
            for (int it = 0; it < Trades().size(); ++it)
            {
                Trades()[it]->StartPath(state[it]);
30          }
        };
```

A second issue is the protection of the member payouts' event times; we must ensure that their DoNode functions are called only at the appropriate times. Thus we implement DoNode with preliminary checking:

```
 —————————————————— CompositeTrade.cpp ——————
     void DoNode
        (const Valuation::UpdateToken_& vls,
         State_* _state,
         NodeValues_& pay_to)
 5   const
     {
        state_t& myS = CoerceComposite(_state);
        assert(myS.Size() == Trades().size());
        const Time_& t = vls.eventTime_;
10      for (int it = 0; it < Trades().size(); ++it)
        {
            if (BinarySearch(eventTimes_[it], t))
                Trades()[it]->DoNode(vls, myS[it], pay_to);
        }
15   }
```

DoDefault is similar, but instead of checking the presence of the event time we check that the default date is before the last event time.

The distinction between collections and sums, shown in the payout by the existence or absence of sumName_, is used when defining the trade value in terms of the streams:

```
 —————————————————— CompositeTrade.cpp ——————
     map<String_, Vector_<pair<String_, double> > >
        CollectWeights() const
     {
```

```
        map<String_, Vector_<pair<String_, double> > > ret;
        for (int it = 0; it < Trades().size(); ++it)
        {
            auto temp = Trades()[it]->StreamWeights();
            for (auto ps = temp.begin();
                 ps != temp.end(); ++ps)
            {
                REQUIRE0(!ret.count(ps->first),
                    "Duplicate trade name in collection");
                PrependToFirst_<String_, double> p
                    (streamPrefixes_[it]);
                ret[ps->first] = Apply(p, ps->second);
            }
        }
        return ret;
    }

    map<String_, Vector_<pair<String_, double> > >
        SumWeights() const
    {
        map<String_, Vector_<pair<String_, double> > > ret;
        auto temp = CollectWeights();
        for (auto p = temp.begin(); p != temp.end(); ++p)
        {
            Append(&ret[sumName_.Value()], p->second);
        }
        return ret;
    }

    map<String_, Vector_<pair<String_, double> > >
        StreamWeights() const
    {
        return sumName_.Known()
            ? SumWeights()
            : CollectWeights();
    }
};
```

The marshalling of backward induction actions proceeds similarly; the prefix for each trade must be prepended to that trade's internal stream names, to create stream names which are seen by the numerical pricer.

Tools for semianalytic pricing of these trades will be discussed in Sec. 14.5.

Chapter 12

Curves

The word *curve* is a term of art, referring to a deterministic function of time used in pricing some set of trades. The prototypical example is the *yield curve*, originally referring to bond yields as a function of maturity but now used for forward Libor rates.

One subtle difference between these usages is worth noting. The bond yields are essentially "raw" market data, depending only on the quoted price in an unambiguous way. Forward rates fitted to some set of quoted instruments, however, are not directly quoted and not fully defined (also see Sec. 7.4). As a rule, we will focus on this latter type of curve.

12.1 Risk

When computing risk, there are two quantities we might demand: the sensitivity to the curve points (forwards) themselves, or to the market quotes used to build the curve. Clearly the relationship between the two is a function of the curve build method, since that method defines how the forwards will respond to a changed quote.

There are three ways to define a curve risk:

- *Store* the curve build method as part of the curve, then apply it to generate bumped curves (for the initial build instruments, or some other specified instruments).
- *Specify* a rebuild method as a required part of the risk specification, thus decoupling the initial build from the risk.
- *Ignore* the build method; construct tractable bumps, such as a constant change to the forwards within a given interval, and compute instrument exposures by inverting the Jacobian of quoted rates to

these bumps.

The "store" method, or the trivial case where only one build method is available, is most commonly used for yield curves; the "ignore" method (also called "bucket hedge") seems most common for ATM volatility curves.

The latter method has the disadvantage that p/l explanation cannot be as precise, because the details of the response to input quotes are not captured. However, the response functions for many common curve builds are of low quality, and it may not be desirable to hew too closely to them. For instance, in yield curve building by linear interpolation of zero-coupon yields from spot, the later rates within an interval between instrument maturities respond more to a bump than do the earlier ones; this will inevitably be reflected in the risk when this method is used to rebuild bumped curves.

In what follows, we choose to ignore the build method; we regard this as superior to storing it. The additional code needed to specify a rebuild method is beyond this volume's scope.

12.2   Libor and Funding

Libor forward curves, and the discount rates that underpin all derivatives pricing, are generally built as a unitary "yield curve." If we consider discounting at Libor, or at a fixed spread to Libor, then we need to build only a single curve – *i.e.*, our yield curve will have only a single functional degree of freedom. But this is a feature of the build method, not of the result; thus we will be careful to avoid making such a promise in our yield curve interface.

The main tasks of a curve are *forecasting* and *discounting*. We separate these in the interface:

```
                         ───── YC.h ─────
   class YieldCurve_ : public Storable_
   {
   public:
      const String_ ccy_;
 5    YieldCurve_(const String_& name, const String_& ccy);
      virtual double DF(long from, long to) const = 0;
      virtual double FwdLibor
         (const PeriodLength_& tenor, long fixing_date)
      const = 0;
10 };
```

This explicitly disavows the dubious operation of changing a curve's currency.

In practice discounting and forecasting are often tied together: *e.g.*, we may write

$$F(t_x) = \delta_t^{-1}\left(\frac{Z(t_s)}{Z(t_m)} - 1\right)$$

where t_x, t_s, t_m are fixing, start and maturity times respectively, δ_t is a daycount fraction and Z is a zero-coupon bond price from a discount curve. This gives forecasts – *e.g.*, of Libor rates – as a function of discounts, for some subset of possible yield curves. The discount curve which yields on-market forecasts of Libor is called "Libor flat."

Often we wish to discount, not at Libor flat, but at some spread to Libor or entity-specific rate. Thus we define a *funding adjustment* which by convention is the ratio of the true discount factor to the Libor-flat discount:

```
──────────────── FundingAdjustment.h ────────────────
class FundingAdjustment_ : public Storable_
{
public:
    FundingAdjustment_(const String_& name);
    virtual double operator()
        (long from_date, long to_date)
        const = 0;
};
```

We support the usual forecasting method with a nonmember utility function

```
──────────────── YCUtils.h ────────────────
double LiborFromDiscounts
    (const YieldCurve_& yc,
    const PeriodLength_& period,
    const FundingAdjustment_* funding,
    long start_date,
    const long* end_date = 0);
```

note that the input date is t_s, not t_x. If t_m is not supplied, the currency defaults for the curve's `ccy_` (see Sec. 8.3) can be used to compute it.

12.3 Build Instruments

Fitting the forward curve to a set of market quotes for standard instruments (the *build instruments*) is a fundamental task. The quotes themselves are generally also available as index fixings; thus we have already created a naming scheme for them, in Sec. 9.1. However, the scheme there cannot take advantage of prior information – that this is an interest rate instrument in a specified currency – and thus yields overly verbose names.

We address this by supporting long instrument names which are the full index names with the `IR[ccy]` dropped from the front: *e.g.*, `SWAP:LIBOR:5Y`. Even these may be considered too long, and we also support an independent set of short names, like `5Y` and `Sep14`, which do not map directly to index names.

Ideally the index parsing and curve instrument parsing would share implementation, but the instrument parser is sufficiently simple that we can relax this constraint. It is sufficient that the same schedule-generation function, which contains all the fragile code, be used for both rate indices and curve instruments.

An instrument supports curve fitting by computing an implied rate, given a candidate curve. Because curve fitting is so frequent, it is worthwhile to optimize this computation; we do this by separating the rate computer from the instrument itself.

```
———————————————— YCInstrument.h ————————————————
class YcInstrument_ : noncopyable
{
public:
    virtual ~YcInstrument_();
    virtual String_ Name() const = 0;
    virtual pair<long, long> TimeSpan() const = 0;

    struct Rate_ : noncopyable
    {
        virtual ~Rate_();
        virtual double operator()
            (const YieldCurve_& yc)
        const = 0;
    };
    virtual Handle_<Rate_> Precompute
        (const Handle_<YcInstrument_>& self,
         const FundingAdjustment_* funding)
    const = 0;
};
```

The `Precompute` function produces the object which will actually compute the rate – all schedule generation, which cannot change across curve scenarios, can be done at this stage. The function's two inputs each correspond to an additional possible optimization. The input `self` will always satisfy `self.get() == this`; thus instruments which need no optimization can derive from `Rate_` and `return self` from `Precompute`. The second input, if non-null, tells the instrument that funding spreads will not vary during the fitting process, so they can be computed once and for all at this stage. This accelerates the most common kinds of curve builds.[1]

12.3.1 *Tenor*

The discount factor implied by compounding successive three-month Libor rates does not, as a rule, agree with that from the corresponding six-month Libor rate. We can address this by generating Libor rates based on a different discount curve for each tenor, or by adding a *tenor spread* to each rate. By convention, "discounting at Libor flat" refers to three-month Libor, regardless of currency; thus the tenor spread will be zero for quarterly rates.

We can efficiently implement such a tenor spread as a two-dimensional interpolant, based on the fixing date[2] and on the Libor length. This length will usually be measured in days, not months: the former will always be available since we need to know the start and end dates, and the latter may be far up the call stack or even (for stubs) nonexistent. To avoid spurious adjustments based on small variations in the number of days of a single Libor tenor, we introduce flat spots in our interpolant around the regular Libor tenors. This means that intermediate Libor rates may not be precisely linearly interpolated, but the resulting errors are very small; and in cases where high precision is important, we will be specifying the interpolation rule in the trade, not in the yield curve (see Sec. 8.5.1).

12.4 Dividend

In equity models with deterministic dividend yields, we create a "dividend curve" which may be either input directly or calibrated to market equity forward prices. This introduces no new challenges, but a couple of points are worth noting:

---

[1] We can go further and supply a partial yield curve, built up to the previous instrument end date, to `Precompute`; but this chains us to bootstrapped curves.

[2] Or on the Libor start date; there is no strong reason to prefer one approach over the other.

- Though the dividend curve may have the same implementation as a rate curve – *e.g.*, the `FundingAdjustment_` curve – it should never be implemented in terms of such a curve. This gratuitously complicates the construction and especially the perturbation of such curves, making sensitivity computations more difficult and restricting implementation changes to either one. Shared functionality should, as always, be factored out into lower-level code.
- Similarly, a two-currency model can be converted into an equity model by converting the equity dividend curve to a yield curve for a fake foreign currency; this may be necessary in the short term but is an obstacle to the longer-term goal of creating component-based models capable of handling both FX and equity.

Production models for equity derivatives should support cash dividends, and even for equity indices the assumption of proportional dividends is undesirable.

12.5   Hazard

To price payouts contingent on default or non-default of some *reference credit*, we must extract the *default intensity* from market instruments. These are generally credit default swaps (*CDS*), or occasionally bonds with issuer risk. We are interested in $Q(T)$, the survival probability to time T in the discount-adjusted measure; and in the default intensity or *hazard rate*, $\lambda(T) \equiv -Q'(T)/Q(T)$. Because we define Q in the appropriate measure, it can immediately be used to compute CDS and bond prices.

Recall from Sec. 8.5.3 that a CDS is described by a fixed coupon *premium leg* with payments c_i at times t_i, $i = 1 \ldots n$; and a *protection leg* of default protection on a notional $N(t)$. For a standard CDS, $N(t) = N_0 \mathbf{1}_{t<t_n}$ is constant up to the CDS maturity, and the value of the protection leg is

$$N_0 \int_0^{t_n} Q(t)Z(t)\lambda(t)dt,$$

where $Z(t)$ is a riskless cash discount factor. The premium leg value, complicated by the payment of accrued interest in the event of default, is

$$\sum_i c_i \left(Q(t_i)Z(t_i) + \int_{t_{i-1}}^{t_i} \frac{u - t_{i-1}}{t_i - t_{i-1}} Q(u)Z(u)\lambda(u)du \right).$$

These integrals are a convenient notation, but we must be aware that they are continuous-time approximations to discontinuous processes; discounting of cash flows is an intrinsically daily event, and default is nearly as discrete.

For valuation purposes, we might choose a coarser discretization to improve performance. It is tempting to use knowledge of the form of λ or even of Z to press this still further: a few exponential integrals, plus some Taylor expansions to handle edge cases, give a feeling of accomplishment. But the more we consider the issues of flexibility and maintenance – driven by desire to build hazard curves that will continue working flawlessly, even as other developers change the fit algorithm or the character of the yield curve – the more merit appears in the brute-force approach.

There is one valuable optimization for the fitting of hazard curves. We will evaluate CDS rates for many candidate solutions, each time with the same $Z(t)$; and, since $\lambda(t)dt = -dQ(t)$ we can eliminate λ in favor of Q everywhere and write the value of a CDS as a linear combination of Q-factors. This precomputed representation can be formed independently of the parametrization chosen for Q or for Z. For a specific fit method, where the parametrization is known, we can optimize this representation further while retaining very high accuracy.

This page intentionally left blank

Chapter 13

Models

To provide numerical pricing, models and their SDE's (see Sec. 10.0.1) must implement the protocols of Ch. 10. Much of this functionality can be provided by the base `SDE_` class[1]; or, if the implementation does not require access to `private` members, in supporting free functions. The latter approach, which separates functionality from the requirement for a particular sort of `this` pointer, is preferred.

Rather than plunge directly into these implementations, we explore the implementation of a specific simple model. In the process we will develop a mix of specific and general functionality; the latter should be easily recognizable, and will mainly be collected in `ModelImp.cpp`.

13.1 Vasicek-Hull-White

The Vasicek model is usually specified by giving an SDE for the short rate, r:

$$dr = \kappa(\mu - r)dt + \sigma dW$$

where κ and σ may be time-dependent, and μ must be strongly time-dependent to reproduce zero-coupon bond prices.

This representation is suboptimal, for two reasons. First, the behavior of μ adds complexity to what is at heart a very simple model; the implementation must concern itself with derivatives of forward rates, which cannot be considered financially meaningful. The second problem is that, by introducing the idea of mean reversion of a short rate, we are exaggerating the role of the short rate, and diverting attention from real tradable quantities.

---

[1] Actually, we prefer to make `SDE_` a pure interface and place shared functionality in `SDEImp_`, from which concrete SDE's are derived.

The mean reversion rate κ determines the values of options, but in a very non-linear way.

These drawbacks can be eliminated, if we instead write

$$r = E[r] + HS, \qquad dS = gdW$$

where g and H are time-dependent, and the expectation is of course in the risk-neutral measure. We then find $\sigma(t) = g(t)H(t)$, $\kappa(t) = -H'(t)/H(t)$. This representation has several advantages:

- Volatilities are linear in the product gH, and option prices nearly so, expediting calibration.
- The HJM volatility $\sigma(t,T) = g(t)H(T)$ is transparently determined.
- Any one-state-variable model is limited to a single mode of yield curve motion; this is simply H.
- The relation of the term structure of g and H to swaption prices is instantly comprehensible.

Thus we end up with a cleaner and marginally more efficient implementation, coupled with a notation which clearly relates the model parameters to option prices.

All the dynamics of the model will depend on a few integrals:

- $\int_{t_-}^{t_+} g^2(u)du$, the variance of S over an interval;
- $B(t,T) \equiv \int_t^T H(u)du$, the sensitivity of bond prices to S;
- $-\int_0^T g^2(t)B(t,T)dt$, which is $\tilde{E}[S]$ in the discount-adjusted measure;
- $\int_{T_-}^{T_+} g^2(t)B^2(t,T_+)dt$, the variance of the innovation in discount factors.

The last integral is required only for Monte Carlo simulation in the risk-neutral measure, which we can avoid; see Sec. 13.1.2.

13.1.1 *Parametrization*

Now, how shall we parametrize g and H? Few-parameter forms should be rejected immediately: the model is already perfectly tractable, and we cannot afford to further reduce its already meager market-fitting capability. In addition, once we have a flexible nonparametric form, we can closely approach any desired parametrization with a special-purpose constructor.

A possible nonparametric form is to make both g and H piecewise-constant functions of time. A generation of quants, trained to believe they are adding value by analytically evaluating complicated integrals, may hold such a simplistic form in contempt; but there is in fact no business justification for anything more complex. Even the smooth exponential decay of H (and growth of g) which leads to a stationary model can be quite well approximated with this method.

Given this parametrization, we store for each interval the values of g and H, and the starting values of the integrals above, accumulated from zero. Since these are used internally by the model, not displayed to calling functions, we place them in **namespace VHWImp**[2]:

```
────────────────────── VHWImp.h ──────────────────────
namespace VHWImp
{
   struct Piece_
   {
5     double g_;
      double h_;
      // integrals to start:
      double B_;
      double varS_;
10    double covDS_;   // will be negative
      double varD ;
   };

   class Vol_ : public Storable_
15   {
      map<Time_, Piece_> vol_;
      void Save(Archive::Dst_& dst) const;
   public:
      Vol_(const map<Time_, Piece_>& vol,
20         const String_& name = String_());

      void Integrate
         (const Time_& from,
          const Time_& to,
25        double* b,
          double* var_s,
          double* cov_d_s,
          double* var_d)
      const;
30
      // allow query of data
      PWC_ G();
      PWC_ H();
```

---

[2]The last two quantities named are a covariance and variance of $\log(D)$, not of D itself, but there is no need to jam this information into their names.

```
         Time_ VolStart() const{return vol_.begin()->first;}
35    };

     Vol_* NewVol
        (const Time_& vol_start,
         const PWC_& g,
40       const PWC_& h);
     }
```

BuildVol lets us construct the volatility from separate PWC_ structures, possibly with different time-dependence; for example, g might be constant if we are only considering options with a single common expiry.

This Storable_ class Vol_ is part of the public type hierarchy (see Sec. 5.2.1), with a type defined by its own mark-up file:

──────────── *VHW_Vol.abstract.storable.if* ────────────
```
BUILDS Handle_<VHWImp::Vol_>
```

We will write Vol_ by writing g and H:

──────────── *VHW_Vol.1.storable.if* ────────────
```
  ISA VHW_Vol
  BUILDS Handle_<VHWImp::Vol_>
  ?STRING name
  PWC g
5    The (time-dependent) volatility of the state S
  PWC H
     The (time-dependent) sensitivity of F(T) to S
```

13.1.2   *Model Contents*

A yield curve, plus a Vol_, is the complete specification of the Vasicek-Hull-White model. Additionally, the model unambiguously obeys a known SDE (see the discussion in Ch. 10). Thus we implement both model interfaces:

──────────── *VHW.cpp* ────────────
```
  class VHW_ : public Model_, public SDE_
  {
     Handle_<YieldCurve_> yc_;
     Handle_<VHWImp::Vol_> vol_;
5  public:
     VHW_(const String_& name,
         const Handle_<YieldCurve_>& yc,
         const VHWImp::Vol_& vol);
```

Note that this is in a source, not a header file; there is no need to gratuitously parade the class definition. As always, we will use factory functions
to create actual instances.

13.2   Interface to Numerical Pricing

We connect the model to our numerical pricers using the model's own state
variables, with no transformation. Monte Carlo simulation is, of course,
agnostic to this choice, since it does not manipulate the state variables
but only passes them to `Stepper_`s and asset models. By displaying a state
variable with highly time-dependent volatility to a PDE, we are committing
to having our PDE solver support a time-dependent grid size; this is not
difficult and adds value for many different models.

A stepper must describe the evolution of S over an arbitrary interval
from t_- to $t_|$, but the no-arbitrage constraints of the model can be satisfied
in several ways.

- We can evolve S in the pure risk-neutral measure, so that the stochastic discount factor D also has an innovation over the interval. This
  requires two random numbers, since D (or $\log D$) is not perfectly correlated with S.
- We can suppress the idiosyncratic innovation in D, using $E[D; S]$ instead; more precisely, $E[D_+; D_-, S_-, S_+]$ where $D_\pm$, $S_\pm$ refer to quantities at time $t_\pm$. This can be thought of as a path integral, over the
  path of S between the event times.
- We can suppress the innovation in D altogether, using instead
  $E[D; D_-, S_-]$. We then adjust the drift of S for the implied change
  of measure. This constructs the *jumping numeraire*.

We want our stepper to consume only a single Gaussian deviate per
step:

```
                        ──── VHW.cpp ────
class VHWStep_ : public ModelStepper_
{
    double sqrtDt_;
    double muS_, sigmaS_;    // de-annualized!
    double a_, bMinus_, bPlus_;
public:
    MonteCarlo::Workspace_* NewWorkspace
        (const MonteCarlo::record_t&)
    const
```

```
10    {
          return 0;
      }
      int NumGaussians() const {return 1;}
      void Step
15        (Vector_<>::const_iterator iid,
          Vector_<>* state,
          MonteCarlo::Workspace_*,
          Random_*,
          double* rolling_df,
20        vector<Handle_<DefaultEvent_> >*)
      {
          const double sMinus = state->front();
          state->front() += muS_ + sigmaS_ * *iid;
          *rolling_df *= exp(a_ - bMinus_ * sMinus
25            - bPlus_ * state->front());
      }
```

Obviously we require $B_+ + B_- = B(T_-, T_+)$. If B_+ (*i.e.*, bEnd_) is zero, this implementation reduces to the jumping numeraire. We prefer

$$B_+ = \int_{T_-}^{T_+} g^2(t) B(t, T_+) dt \Big/ \int_{T_-}^{T_+} g^2(t) dt,$$

i.e., we reproduce the covariance of S with $\log D$; thus we can set $\mu_S = 0$. Now the joint distribution of $S_\pm$ is known, and we derive a to fit the yield curve, finding

$$a = \log\left(\frac{Z_+}{Z_-}\right) + (B_- + B_+)\tilde{E}[S_-] - \frac{B_-(B_- + 2B_+)}{2}\mathrm{Var}[S_-] - \frac{B_+^2}{2}\mathrm{Var}[S_+],$$

where $Z_\pm$ denotes the discount factor to $T_\pm$ and $\tilde{E}$ is an expectation in the discount-adjusted measure to time T_-. Note that the calculation of $\tilde{E}[S]$ is already supplied by the parametrization described in Sec. 13.1.1.

Next we proceed to form the quantities needed for (backward induction) PDE pricing. We have postulated that S is a martingale, so we need no advection term; the diffusion term is sigmaS_ / sqrtDt_.[3] Only the discounting is S-dependent, with the rate

$$(B_- + B_+)S - \frac{a}{\Delta_t}.$$

Thus only the discounting function requires its own implementation class.

---

[3]If Step is used at all, it will be called far more frequently than the PDE coefficient calculators; thus we favor it in our optimization.

```
 ──────────── VHW.cpp ────────────
     PDE::ScalarCoeff_* DiscountCoeff() const
     {
        struct Mine_ : PDE::ScalarCoeff_
        {
5           x_dep_t xDep_;
            double c0_, c1_;
            Mine_(double c0, double c1) : c0_(c0), c1_(c1)
            {
                xDep_.set(0);
10          }
            void Value(const Vector_<>& x,
                double* value)
            const
            {
15              assert(x.size() == 1);    // just S
                *value = c0_ + c1_ * x[0];
            }
            x_dep_t XDependence() const {return xDep_;}
        };
20      const double dt = Square(sqrtDt_);
        return new Mine_(a_ / dt, (bPlus_ + bMinus_) / dt);
     }
     PDE::VectorCoeff_* AdvectionCoeff() const
     {
25      return PDE::NewConstCoeff(Vec1(0.0));
     }
     PDE::MatrixCoeff_* DiffusionCoeff() const
     {
        SquareMatrix_<> coeff(1);
30      coeff(0, 0) = 0.5 * Square(sigmaS_ / sqrtDt_);
        return PDE::NewConstCoeff(coeff);
     }
};
```

We do not use the **StepAccumulator_** in creating steppers, but we still must provide envelope information for PDE solvers.[4]

```
 ──────────── VHW.cpp ────────────
struct VHWAccumulator_ : StepAccumulator_
{
    Time_ volStart_;
    const VHWImp::Vol_& vols_;
5   Vector_<pair<double, double> > Envelope
```

---

[4]Here we construct the envelope in the discount-adjusted measure. It might be considered safer to give the upper bound in the risk-neutral measure (remember $\tilde{E}[S] \leq 0$).

```
          (const Time_& t,
           double num_sigma)
      const
      {
10        double varS, eS;
          vols_.Integrate(volStart_, t, &varS, 0, &eS, 0);
          const double halfWidth = sqrt(varS) * num_sigma;
          return Vec1(make_pair(eS-halfWidth, eS+halfWidth));
      }
15 };
```

13.3   Interface to Valuation Requests

Upon receipt of a completed value request, we must create an asset model which will honor those requests by computing the desired market quantities, as a function of the model state S. The SDE_ class will provide a template method to create an updater, based on calls to the model to produce a value for a single index.

In this model we assume that changes in the state drive all discounts, including those implied from Libor rates, in the same way, so that funding adjustments are deterministic. A Libor rate starting from t_d can used to construct a discount factor, equal to $1+\delta L$ where δ is the daycount fraction, over the interval from t_d to the Libor maturity t_m. Thus we have

$$1 + \delta L = Ae^{B(t_d,t_m)S},$$

where A is determined by the necessity of repricing Libor in the discount-adjusted measure to t_m:

$$1 + \delta F = A\tilde{E}_{t_m}\left[e^{B(t_d,t_m)S}\right]$$

where F is the forward Libor rate from the yield curve. Thus we know our final code must rely on a supporting function[5]

──────────────────── ***VHW.cpp*** ────────────────────
```
SDEImp::UpdateOne_* NewLiborUpdater
   (_ENV, const YieldCurve_& yc,
   const VHWImp::Vol_& vol,
   const Time_& event_time,
5  long start_date,
   const PeriodLength_& tenor)
{
```

---

[5]The only subtle feature of this function is the choice of arguments to operator(), which we are about to explain.

```
     struct Mine_ : SDEImp::UpdateOne_
     {
10      double A_, B_, dct_;
        Mine_(double A, double B, double dct)
            : A_(A), B_(B), dct_(dct) {}
        double operator()(Vector_<>::const_iterator& state,
            const Valuation::UpdateToken_& prior) const
15      { return (A_ * exp(B_ * state[0]) - 1.0) / dct_; }
     };

     const double F = yc.FwdLibor(tenor, start_date);
     const long tm = Date::NominalMaturity
20          (start_date, tenor, yc.ccy_);
     const double dct = Ccy::Conventions::LiborDayBasis()
            (_env, yc.ccy_)(start_date, tm);
     double B, ES, varS;
     vol.Integrate(start_date, tm, &B, 0, 0, 0);
25   vol.Integrate(vol.VolStart(), tm, 0, &varS, &ES, 0);
     const double A = (1 + dct * F)
            * exp(-B * (ES + 0.5 * B * varS));
     return new Mine_(A, B, dct);
}
```

This function could be called from a member function like
VHW_::NewUpdater whenever a Libor rate is needed. However, the above
code embeds some important limitations. It assumes complete self-
knowledge on the part of the model – to compute a Libor, we are given
the **state** and nothing else. This is theoretically correct, and has no ill effects in this simple model; but when we proceed to component-based models
we must permit the components to *make requests of each other* to separate
their implementations. For instance, an equity component will fulfill a request for a forward equity price by in turn requesting a discount factor –
thus avoiding the need to know about the process by which discounts are
computed from the interest-rate state.

The need for updaters sometimes to see the results of earlier updates
dictates the interface

```
───────────── SDEImp.h ─────────────
namespace SDEImp
{
   class UpdateOne_ : noncopyable
   {
5  public:
      virtual ~UpdateOne_();
```

```
      virtual double operator()
          (Vector_<>::const_iterator& state,
            const Valuation::UpdateToken_& prior)
10    const = 0;
    };
}
```

We pass the `state` as an iterator, rather than a whole vector, to better support composite models built of independent components: see Sec. 13.7.

Individual updates are weakly ordered by their possible dependency relationships. The requests which the model receives from the trade are ordered by their `IndexKey_`, but this is not relevant to the model. We must produce updaters which are called in the correct sequence, so that the needed `prior` information is always available for a given update. The dependency tracking for this is handled inside `SDEImp_`, so the interface to the asset can be at the highest level:

SDE.h

```
namespace SDEImp
{
    class Update_ : noncopyable
    {
5   public:
        virtual ~Update_();
        virtual void operator()
            (Vector_<>* vals,
              const Vector_<>& state,
10            const Valuation::UpdateToken_& pass)
        const;
    };
}
```

The `operator()` will be called with `pass.begin_` == `vals->begin()`; thus the `Update_` can populate `vals` while also allowing `UpdateOne_` functions to access it as their `prior`. The `SDEImp_` is responsible for ensuring that this unsafe access is always in the correct order.

Obviously there is an `Update_` for each event time; creation of these requires non-`const` access to the `ValueRequest_` being satisfied. We provide `SDEImp_` with the member function

SDE.h

```
    SDEImp::Update_* NewUpdate
        (_ENV, ValueRequestImp_& requests,
          const Time_& event_time)
```

```
      const;
```

ValueRequestImp_, which is the base type of all concrete **ValueRequest_s**, is discussed in Sec. 10.6.1. Here we will see the need for its **AtTime** query.

To implement **Update_** with a collection of **UpdateOne_**, we need to store the single updates and the destinations for their outputs.

```
————————————————— SDEImp.cpp —————————
   struct UpdateImp_ : SDEImp::Update_
   {
      Vector_<pair<int, Handle_<SDEImp::UpdateOne_> > >
            updates_;
 5    void operator()
         (Vector_<>* vals,
          const Vector_<>& state,
          const Valuation::UpdateToken_& pass)
      {
10       assert(&pass[0] == &vals->front());
         for (auto pv = updates_.begin();
                 pv != updates_.end(); ++pv)
         {
            (*vals)[pv->first]
15             = (*pv->second)(state.begin(), pass);
         }
      }
   };
```

This will only work if we ensure that the **updates_** are executed in order. To track dependencies, the **SDEImp_** must mediate access to the **ValueRequest_**; thus we create another object to pass to derived **SDE_s** making single-index evaluators.

```
———————————————————— SDE.h ———————————
   class RequestAtTime_
   {
      ValueRequestImp_& request_;
      map<IndexKey_, Valuation::address_t> all_;
 5    bool stale_;            // if all_ needs updating
      set<IndexKey_> done_;
      vector<Handle_<Index_> > wait_;    // priors

      friend class SDEImp_;
10    typedef Handle_<SDEImp::UpdateOne_> update_t;
      RequestAtTime_(ValueRequestImp_& r, const Time_& t)
      : request_(r), t_(t), all_(r.AtTime(t)), stale_(false)
      {    }
```

```
      void Refresh()
15    {
          if (stale_)
              all_ = request_.AtTime(t_);
          stale_ = false;
      }
20    void Push(Vector_<pair<int, update_t> >* updates,
          const IndexKey_& index, const update_t& update)
      {
          updates->push_back(make_pair(all_[index], update));
          done_.insert(index);
25    }
  public:
      const Time_ t_;
      Valuation::address_t operator()(const Index_& index);
  };
```

This `operator()` will obtain a promised fixing location from its `request_`,
and will also note whether the request is a new one (whose update must
therefore precede the update being formed).

```
————————————————— SDE.h ——————————————
Valuation::address_t RequestAtTime_::operator()
      (const Index_& index)
  {
      // make the key we need
5     IndexKey_ key(index);
      if (!done_.count(key))
      {   // we don't have this update yet
          wait_.push_back(key.val_);
      }
10    if (!all_.count(key))
      {
          // we haven't seen this request yet
          stale_ = true;
      }
15    return request_.Fixing(t_, index);
  }
```

We store the dependencies of some updates on others as *continuations* which
associate a deferred task with those that must precede it.

```
————————————————— SDEimp.cpp ——————————————
namespace
{
    struct Continuation_
    {
```

```
 5      Handle_<Index_> index_;
        Handle_<SDEImp::UpdateOne_> update_;
        Vector_<Handle_<Index_> > wait_;
        Continuation_(const Handle_<Index_>& index,
           const Handle_<SDEImp::UpdateOne_>& update,
10         const Vector_<Handle_<Index_> >& wait)
         : index_(index), update_(update), wait_(wait) {}
      };
   }    // leave local namespace
```

Now we can finally implement **NewUpdate**:

```
——————————— SDEimp.cpp ———————————
SDEImp::Update_* SDEImp_::NewUpdate
   (_ENV, ValueRequestImp_& requests,
   const Time_& event_time)
const
 5 {
      auto_ptr<UpdateImp_> retval(new UpdateImp_);
      stack<Continuation_> tbc;
      RequestAtTime_ toDo(requests, event_time);
      for ( ; ; )
10    {
         Handle_<SDEImp::UpdateOne_> update;
         Handle_<Index_> next;
         if (tbc.empty())
         {   // no continuations
15          toDo.Refresh();
            if (toDo.all_.size() == toDo.done_.size())
               break;
            next = FindAnother(toDo.all_, toDo.done_);
         }
20       else if (tbc.top().wait_.empty())
         {   // now ready for this one
            next = tbc.top().index_;
            update = tbc.top().update_;
            tbc.pop();
25       }
         else
         {   // still clearing the decks
            next = tbc.top().wait_.back();
            tbc.top().wait_.pop_back();
30       }

         // get an update, and see whether we have to wait
         assert(toDo.wait_.empty());
```

```
          if (update.Empty())
35            update.reset(NewUpdateOne(_env, toDo, *next));

          if (toDo.wait_.empty())
          {   // we can do it now
              toDo.Push(&retval->updates_, next, update);
40        }
          else
          {   // put it on hold, do others first
              tbc.push(Continuation_(next, update, toDo.wait_));
              assert(!IsCircular(tbc));
45            toDo.wait_.clear();
          }
      }
      return retval.release();
}
```

This code is complex, but its general outline should be clear enough. We start by choosing an index in `all_` but not `done_` – the routine `FindAnother` accomplishes this. We create an update for it, with `toDo` to tell us what other updates it depends on. Since the call to `Push` updates `toDo.done_` and `retval->updates_` are together, they are reliably synchronized. If the new update depends on updates which are not `done_`, then we postpone it – pushing in onto a stack of `Continuations_` on top of those updates which depend on it – while we work through those dependencies. When an index is taken out of `tbc.back().wait_`, it next appears either in `done_` or as a new entry in `tbc` awaiting its own turn; either way, it is sure to be processed before `tbc.back().index_`, whose update depends on it.

This routine will fail utterly – with a stack overflow – if there is a circular dependency among the updates. We can detect this by testing that the `index_` entries of each stored `Continuation_` are unique; this is the function of `IsCircular` in the displayed code.[6] This must arise from a grievous coding error, so an `assert` is appropriate.

13.3.1 *Index Paths*

In the absence of index path requests, an `Update_` for each of the trade's event times is sufficient to form an `Asset_`. To form index paths, we need to add to `SDEImp_` a new member

---

[6] Actually, `IsCircular` must examine the whole `stack` of continuations, so it will be grossly inefficient unless we change `tbc` to a vector.

```
───────────── SDEImp.h ─────────────
IndexPath_* NewIndexPath
    (const Index_& index,
     const Vector_<Time_>& index_dates,
     ValueRequest_& request);
```

The input **index_dates** are the trade's event dates, excluding those beyond the last date on which the **IndexPath_** is used.[7] The path will naturally depend on the fixings at event dates, which we will obtain from the **ValueRequest_**; thus the index-path updates are *created* before but *executed* after the fixing updates.

Information about path behaviour between event dates – *e.g.*, volatilities for a Brownian bridge – will generally not require cross-index updates. Composite indices can compute such quantities only if the model describes the joint behavior of indices between event dates; this is beyond the scope of this volume.

13.3.2 *Efficiency*

When we are working through the continuation stack in **NewUpdate**, we know we have not exhausted the requested indices; thus we **Refresh** only when the stack is empty, when we ask the **request** for more tasks. For simple models without the possibility of dependencies between requests, **tbc** will always be empty and the flow of control inside the loop reduces to

```
    if (toDo.all_.size() == toDo.done_.size())
        break;
    IndexKey_ next = FindAnother(toDo.all_, toDo.done_);
    Handle_<SDEImp_::UpdateOne_> update
            (NewUpdate(toDo, next)),
    toDo.Push(&retval->updates_, next, update);
```

The completeness check and **NewUpdate** take $O(1)$ time, regardless of the number of indices being updated; and **Push** takes $O(\log n)$ time because of the call to **set::insert**. A naive implementation of **FindAnother** would walk through **all_** and **done_** together, taking $O(n)$ time for each call (thus $O(n^2)$ total time). We can avoid this by writing

---

[7]There is a very slight inefficiency here, in that we require the construction of the vector of index_dates rather than just passing iterators.

```
─────────────────────── SDEimp.cpp ───────────────────────
  Handle_<Index_> FindAnother
      (const map<IndexKey_, Valuation::address_t>& all,
       const set<IndexKey_>& done)
  {
5     assert(all.size() > done.size());
      auto pa = done.empty()
          ? all.begin()
          : all.upper_bound(*done.rbegin());
      if (pa == all.end())
10        pa = all.begin();      // fast method failed
      set<IndexKey_>::const_iterator pd = done.begin();
      while (pd != done.end() && pa->first == *pd)
      {
          ++pa, ++pd;
15        assert(pa != all.end());
      }
      return pa->first.val_;
  }
```

which runs in $O(\log n)$ time in the simple case, and is usually (though not guaranteed) fast in more complex cases.

Another optimization has to do with code caching. If several updates – *e.g.*, of equities within a basket – are evaluated by the same code, then we can reduce cache faults and improve performance by grouping those updates together. This requires a functor to inspect an index and determine whether its evaluation should be brought forward:

```
─────────────────────── SDEImp.h ───────────────────────
  struct IndexIsSimilar_ : noncopyable
  {
      virtual bool operator()(const Index_& index) const = 0;
  };
```

Then we add

- A local variable, `scoped_ptr<IndexIsSimilar_> sim`, alongside `monitor` and `tbc` above;
- An additional argument, `scoped_ptr<IndexIsSimilar_>* sim`, to the single-index version of `NewUpdate` – we will send `&sim` to this routine;
- An additional argument, `const IndexIsSimilar_* sim`, to `FindAnother` – we will send `sim.get()`.

Each call to `NewUpdate` has the option to reset `sim` to a new test function; when a non-`NULL` test function is provided, `FindAnother` must attempt to return a preferred index, but also account for the possibility that no such index exists. We can accomplish this by creating a circular iterator which loops around `all`, starting from the initial `pa` above; and saving a fallback value to be returned if `(*sim)(x)` is `false` for all keys `x`. This code can no longer be expected to run in $O(\log n)$ time, so it can make the setup phase expensive for a large number of indices unless they are almost all similar.

In our examples of `NewUpdate`, we do not implement this optimization.

13.3.3 *Back to Libor*

The VHW implements specific updaters for any indices it knows how to handle; to accomplish this, it must inspect the index. We can use `dynamic_cast`:

```
                         ─── VHW.cpp ───
    SDEImp::UpdateOne_* VHW_::NewUpdateOne
        (_ENV, const RequestAtTime_& t,
        const Index_& index)
    const
  5 {
        if (DYN_PTR(ir, const IndexIr_, &index))
        {
            Require(_env, ir->ccy_ == yc_->ccy_,
                "Non-domestic rate index requested");
 10         const long start = ir->StartDate(t.t_);
            if (IsLiborTenor(ir->tenor_))
                return NewLiborUpdater(*yc_, *vol_, t.t_, *ir);
            if (IsSwapTenor(ir->tenor_))
                return SDEImp::NewSwapUpdater
 15                 (*yc_, *vol_, t, *ir);
            Throw(_env, "Unrecognized interest rate tenor");
        }
        if (DYN_PTR(df, const IndexDf_, &index))
        {
 20         const long mat = df->Maturity(t.t_);
            return NewDfUpdater
                (*yc_, *vol_, t.t_, df->Maturity(t.t_));
        }
        Throw(_env, "VHW model can't simulate non-IR indices");
 25 }
```

Here `NewLiborUpdater` and `NewSwapUpdater` are local free functions which do whatever precomputation is possible. The definition of `NewLiborUpdater` given above needs to change only to accomodate the new definition of `UpdateOne_::operator()`.

In the code above, we pass `t` rather than just `t.t_` to `NewSwapUpdater`. This allows the swap updater to request that other index values – Libor rates and discount factors – be available when it is invoked. This decouples the swap rate computation from the VHW model, so this functionality can be provided to all SDE's.

For the VHW model, we can use this default implementation, or can create our own optimized swap updater (*e.g.*, using cubic-spline interpolation to get the swap rate as a function of S).

13.4 Cox-Ingersoll-Ross

The Cox-Ingersoll-Ross (CIR) model relies on a short rate dynamics of the form

$$dr = \kappa(\mu - r)dt + \sigma\sqrt{r}dW.$$

This model has some disadvantages compared to the VHW model we have just discussed:

- Its parametrization cannot easily be recast into the gH-form, because rates remain positive only when $2\kappa\mu > \sigma^2$. Thus a region of negative κ, or an increase in H, permits negative rates from which the model cannot recover.
- The bond pricing formula is still of the form $P = Ae^{-Br}$, but the computation of the coefficients is substantially more complicated, especially when κ and σ are not constant.
- Implementation of an arbitrage-free Monte Carlo stepper is quite complex (as it is for the Heston model of a single asset price), and the stepper cannot be expected to be highly efficient.
- The $\beta = \frac{1}{2}$ behavior is hard-wired into the model, just as thoroughly as $\beta = 0$ in the VHW model.

On the other hand, it is plausible that $\beta = \frac{1}{2}$ is a better choice, more likely to be close to the market implied elasticity.

For these reasons, we probably would not trouble to implement this model. Should we choose to do so, however, the path is reasonably clear.

We will make κ and σ piecewise constant, so that the coefficients A and B of the bond pricing formulas will remain available in closed form. Bond options can then also be priced in closed form, and the PDE coefficients are readily available. The Monte Carlo step requires some degree of approximation.

13.5 Black-Karasinski

The Black-Karasinski (BK) model, which extends the earlier Black-Derman-Toy model with the most flexible one-state-variable dynamics, is usually written

$$d\log r_t = \kappa(t)(\theta(t) - \log r_t)dt + \sigma(t)dW_t$$

where, as before, θ must be time-dependent to fit the yield curve, and κ and σ may be time-dependent to allow a more flexible volatility structure. Note that this is just the VIIW dynamics with r_t replaced by $\log r_t$. We can rewrite this model in the form

$$dS_t = y(t)dW_t, \qquad \log r_t - E[\log r] + H(t)S_t.$$

While this captures the features of the "classical" BK model, the short rate elasticity is fixed at one, so there is no gain in flexibility to compensate us for the inevitable loss of analytic tractability.

For a more flexible model form, we rewrite the relation between r and S as

$$r_t = \bar{r}(t)e^{H(t)S_t}$$

where $\bar{r}$, like $E[\log r]$, must be fitted numerically to the yield curve. This shows that the lognormal behavior of the BK model is only one choice among many, and can be replaced by a more general functional form. Models based on this idea are often called *generalized Brownian motion* (GBM) models.

What market features should we attempt to model with this new freedom? We can certainly use it to model skew, by giving r the *shifted lognormal* dynamics

$$r_t = \Big(\bar{r}(t) + \Delta(t)\Big)e^{H(t)S_t} - \Delta(t)$$

for a (possibly time-dependent) shift Δ. Adding a quadratic dependence on S will fatten the tails of the distribution and increase the value of out-of-the-money options; but at some point $r(S)$ will no longer be monotonically increasing and the implied yield curve motions will become markedly un-realistic. Similarly, by creating a rich nonparametric dependence of r on S we can hope to fit option prices across a range of strikes, but the resulting models are often plagued by outlandish parameter values and can be used only with the most extreme care.

The pricing code is largely independent of the parametrization, so we do not need to commit firmly to a single choice.

13.5.1 *Forward Induction PDE Sweep*

One feature we demand of any parametrization is a parameter – here called $\bar{r}$ – which can be tuned to match the yield curve. We may choose to fit this parameter as part of the pricing process, using forward induction out to the trade's last maturity date, or during the construction of the model, with the last date provided as part of the model specification. We choose the former approach, performing the fit as part of the creation of an SDE_ from the model (see Sec. 10.0.1).

The PDE solver will propagate S_t, and the mapping $r(S)$ will form the ScalarCoeff_ used for discounting. Fitting $\bar{r}$ involves a numerical search, at each time step, to match the observed discount factor $Z(T)$; this fit can be made much more efficient (converging in 2-3 steps) if we are able to compute $dr/d\bar{r}$ at each S.

This leads us to an interface for the abstract mapping:

```
                        ──────── BKImp.h ────────
   namespace BK
   {
      class Mapping_ : public Storable_
      {
5     public:
         double rBar_;
         virtual double R(double S) const = 0;
         virtual double dRdS(double S) const = 0;
      };
10  }
```

Note that each Mapping_ is valid for only a single time step; but if a para-metric structure is shared across time steps, we can store it in a Handle_ shared across Mapping_s.

13.6 Single Equity with Local Vol

Models with deterministic but price-dependent "local" volatility are quite common in equities modeling, partly because they promise to decompose basket options on several equities by computing sensitivity to single-equity options at an appropriate range of strikes. For heavily path-dependent products – most specifically, for those with strongly *path-dependent gamma* – the assumption of deterministic vols is not justified.

When choosing the representation of such a model, we must immediately decide whether to parametrize the local vols or the implied term vols, from which we can rapidly infer local vols using the celebrated Dupire formula. The latter approach, which makes vanilla option prices immediately available, is more common in practice; but we find it distasteful, since it obscures rather than reveals the actual dynamics used in pricing. Here we will parametrize the local vols directly.

For the present, we assume that interest rates are deterministic; we will return to this issue in the next section. Also, we consider only a single equity process, though most structured equity derivatives involve several underlyings.

We first define an abstraction of a local volatility surface:

```
────────────────────── LVSurface.h ──────────────────────
    class LVSurface_ : public Storable_
    {
    public:
        virtual Time_ VolStartTime() const = 0;
5       virtual double LocalVol(const Time_& t, double s)
            const = 0;
        virtual double IntervalVol(const Time_& t_minus,
            const Time_& t_plus, double s) const;
        // ...
10  };
```

The `IntervalVol` function is not pure virtual, because we can reasonably implement it with a set of queries to `LocalVol`. The following recipe works reasonably:

(1) Let $\sigma_\pm$ be the local vol at $(t_\pm, s)$.

(2) Let $\bar{t} = \frac{t_+ + t_-}{2}$ and $\delta_t = t_+ - t_-$.

(3) Let $h = \frac{\sigma_+ + \sigma_-}{(\sigma_+ + \sigma_-)} \sqrt{\delta_t}$, the central Brownian bridge width.

(4) Let σ_u be the local vol at $(\bar{t}, se^h)$ and σ_d at $(\bar{t}, se^{-h})$.

(5) The annualized step variance is about $V \equiv \frac{(\sigma_-^2 + \sigma_+^2 + 2\sigma_u^2 + 2\sigma_d^2)}{6}$; use $\sqrt{V\delta_t}$

as the step vol.

13.6.1 *Interpolated Vol*

Our favored parametrization will be based on interpolation in a central region, and CEV-like extrapolation at extreme values of S:

```
──────────── LVInterp.cpp ────────────
class LVInterp_ : public LVSurface_
{
    Time_ volStart_;
    Handle_<Interp2_> vals_;
5   Interp::OfTime_ loEdge_, loBeta_;
    Interp::OfTime_ hiEdge_, hiBeta_;
```

The main member, `vals_`, defines the local (lognormal) volatility in some central region. For extreme strikes, we switch to a CEV dynamics; the `loEdge_` and `hiEdge_` are the bounds of the central region, and the corresponding β's define the extrapolation. We allow all these quantities to be time-dependent, though it makes little difference in practice.

This defines the local vol:

```
──────────── LVSurface.h ────────────
public:
    Time_ VolStartTime() const {return volStart_;}
    double LocalVol(const Time_& t, double s) const
    {
5       const double lo = loEdge_(t);
        if (s < lo)
            return LocalVol(t, lo) *
                pow(s/lo, loBeta_(t)-1.0);
        const double hi = hiEdge_(t);
10      if (s > hi)
            return LocalVol(t, hi) *
                pow(s/hi, hiBeta_(t)-1.0);
        return vals_->Get(t.DDate(), s);
    }
15 };
```

13.6.2 *Derivation from Implied Vol*

Despite our aesthetic objections, local vols are often hidden behind a global parametrization of implied vols, and derived as needed using the Dupire formula. The derivation of local vols around the time of a discrete dividend payment, required for a production-quality implementation, requires great care. It is also likely that an override of `IntervalVol` can produce superior results to the purely local default implementation, since it can use the global implied vol information.

13.6.3   *Model and SDE*

Now let us wrap these dynamics in an SDE to support numerical pricing. For simplicity, we will consider only proportional dividend yields, parametrized using a `PWC_` object. Again, the `Model_` and `SDE_` can share an identity:

```
————————————————— LVModel.cpp —————————————————
class LVModel_ : public Model_, public SDE_
{
    double s0_;
    Handle_<YieldCurve_> yc_;
    Handle_<PWC_> divs_;
    Handle_<LVSurface_> vols_;
    // ...
```

The stepper is similar to that of the VHW model (Sec. 13.2), but naturally it must report a space-dependence of the diffusion coefficient. That can be computed using repeated calls to `LVSurface_::IntervalVol`; this is not so bad for PDE's, but to support a production Monte Carlo we will find it necessary to do some precomputation. For instance, the `IntervalVol` could be precomputed at a hundred or so points covering the envelope, and splined to find the appropriate vol for a given simulation path.

The computation of the envelope is also more involved, because of the lack of global information. It is actually best to introduce this as a new virtual member in `LVSurface_`:

```
————————————————— LVSurface.h —————————————————
virtual StepAccumulator_* NewAccumulator() const = 0;
virtual pair<double, double> UpdateEnvelope
    (StepAccumulator_* accumulator,
     const Time_& t,
     double num_sigma)
    const = 0;
```

The call to `NewAccumulator` creates a derived class, visible only to the vol surface, containing information about the envelope so far. Then each call to `UpdateEnvelope` advances the `accumulator` to the new time t, and returns an envelope computed from it. For a Dupire-derived surface no accumulator is needed, and `NewAccumulator` can return 0; for our interpolated local vols, we use a weighted average of vols in the central region and at the existing envelope edge to widen the envelope.

13.7   A Simple Hybrid Model

Now let us combine this local-vol model with our Vasicek-Hull-White model. At the same time, we will consider modeling with more than one equity. For this, we wish to reuse the code of the single-process models we have already created.

Thus we separate the hybrid model into *components* – the VHW and local vol processes – and *correlations*. The latter will be part of the top-level model specification. We also introduce the concept of *polling*, where for a given request – *e.g.*, an index for which a fixing must be generated – we query the components to find one which can honor the request.

The model should store only a single yield curve – otherwise the process of updating the curve, *e.g.* for a curve delta computation, is gratuitously complicated. Thus we begin by rewriting our local vol model, collecting the description of the equity process into a single component:

```
———————————— LVModel.cpp ————————————
class LVModel_ : public Model_
{
   Handle_<YieldCurve_> yc_;
   Handle_<LVComponent_> equity_;
public:
   LVModel_(const String_& name,
      const Handle_<YieldCurve_>& yc,
      const Handle_<LVComponent_>& eq);
   // ...
```

The component, and the asset models it creates, can no longer assume that the **state** passed to them is entirely theirs. In constructing an asset model, we must supply a *state offset* which will determine where the component's state is stored inside the whole model state.[8]

Components, like **SDE_s**, must make asset updaters, and will receive a **RequestAtTime_** which they can use to communicate with other components. Thus, once the step is taken and the new **state** is available, our

---

[8]This offers no protection against components reading or writing out of range. Enforcing such protection through a system of types is possible but expensive at runtime – remember that asset pricing is a code hotspot. In Sec. 13.7.2 we will show a useful protective mechanism.

existing machinery suffices to support component models. It is best, when creating the updaters, to be sure that the result is unique – that the model does not somehow contain two components which each think they know how to compute a fixing for some given index.

The sharing of information during stepping is harder. For instance, an equity stepper needs to know the discount rate over the step interval, which supplies its drift term. Attempting to reuse the dependency mechanism of NewUpdate would come at a high price in both maintainability – as we are mixing step and post-step code – and efficiency. A more specialized tool is needed.

We observe that there are only a few ways in which the output of one stepper can be input to another: discount rates, FX vols (for quanto drift), and rate vols (for drift of spread rates). The last is beyond the scope of this volume; we will discuss mechanisms for only the first two.

To ensure that discount factors for the step are available before they are needed, we bring rate-stepping components to the front and execute them first. But foreign rates will require a quanto shift, which depends on knowledge of the FX vol. Thus we need to ensure that the ordering is

(1) The domestic rate stepper.
(2) All FX steppers.
(3) All foreign rate steppers.
(4) An additional phase, where the FX steppers are given a discount factor to complete their update.
(5) All other steppers.

After all this, a component sub-stepper requires several inputs not visible to the top-level Monte Carlo:

―――――――――――――― *ComponentStep.h* ――――――――――――――
```
class SubStepper_ : noncopyable
{
public:
    virtual ~SubStepper_();
    virtual MonteCarlo::Workspace_* NewWorkspace
        (const MonteCarlo::record_t&)
    const = 0;
    virtual int NumGaussians() const = 0;
    virtual void Step
        (Vector_<>::const_iterator iid,
        Vector_<>::iterator state,
         MonteCarlo::Workspace_* work,
        ComponentDf_* dfs,
```

```
                 ComponentQVols_* quantos,
15               Random_* extras,
                 vector<Handle_<DefaultEvent_> >* defaults)
       const = 0;
       virtual void ApplyDf(const ComponentDf_& dfs) const {}
   };
```

The structures `ComponentDf_` and `ComponentQVols_` contain discount factors and FX lognormal vols over the step period, respectively; they are read by some components and written by others. Steppers for FX components, which will be called before the `dfs` can be fully populated, are expected to override `ApplyDf` and use it to update the FX state once the necessary discounts are known.

The most natural implementation of `ComponentDf_` would be a `map<String_, double>` but this relies on inefficient string comparisons; it is better to reuse the `CreditId_` in Sec. 10.5 to translate the currencies (held by the components) into integer locations (held by the steppers) within a simple `Vector_<>`.

The top-level *master stepper* synchronizes the calls to `SubStepper_s` and communicates a single rolling discount back to the numerical method. It always has a workspace of its own, so that space for the `dfs` and `quantos` need not be reallocated for every path.

13.7.1 *The Case for Components*

Re-implementing a model in this way involves intensive work, and careful setup of generic mechanisms to do simple jobs without too much inefficiency. Given the comparative ease of simply dispatching all the information manually – creating a specialized `LVPlusVHW_` model and hand-writing an optimal stepper for it – it is natural to ask why one should go to the trouble of isolating components.

The answer is that development is a continuous process, and we can make the most progress by opening the door to small successes on one front at a time. Upgrading a single specialized model is a wide-ranging task, and we have nothing to show until it is fully done. But once the connections between components are formed generically, and the role of each individual component is made explicit, then upgrading one kind of component without needing to recode the entire model becomes a faster path to the creation of valuable models yielding a competitive advantage.

A further reason is that a risk measurement system is only as strong as its weakest link; it is fruitless to control interest rate skew exposures in a Bermudan book if we also have a PRDC book where the same exposures cannot even be measured. Component-based models let us refine a process which will give an interesting measure of our portfolio's exposure, and then deploy it to measure *all* the derivative trades with relevant risks.

13.7.2   *State Bounds Checks*

The model queries each of its components for the state size, and builds a composite state joining each component state. By placing internal padding inside this composite state, we can detect most range errors. At the outer (model) level, we form the state including extra elements:

```
                          LVHWModel.cpp
  #ifdef debug
  static const Vector_<> STATE_PADDING(4, DA::NAN);
  #else
  static const Vector_<> STATE_PADDING;
5 #endif

  Vector_<pair<double, double> > LVHWAccumulator_::Envelope
     (const Time_& t,
      double num_sigma)
10 {
     static const Vector_<pair<double, double> > PADDING
           = Zip(STATE_PADDING, STATE_PADDING);

     Vector_<pair<double, double> > retval
15          = ir_->Envelope(t, num_sigma);
     for (auto pe = eq_.begin(); pe < eq_.end(); ++pe)
     {
         Append(&retval, PADDING);
         Append(&retval, (*pe)->Envelope(t, num_sigma));
20   }
     return retval;
  }
```

Write errors are checked at the model level, by asserting that the padding elements remain equal to `DA::NAN` (or another extreme constant of our choice). Read errors are checked at the component and asset level, by asserting that a state variable does not have this value.

This protection cannot be used in PDE pricing, since it expands the state with extra variables that have no dynamics. This is not much of a drawback in practice; most hybrid models already have too many state variables for any feasible PDE, and this kind of check is most useful when individual components have many state variables – otherwise range errors are not a threat.

Chapter 14

Semianalytic Pricers

All our discussion so far has been centered on generic numerical pricing, where models and trades can communicate at arm's length. We now turn to specialized *semianalytic methods*, which use knowledge of both the trade and model to rapidly obtain the price, or a close approximation thereto.

For example, a lognormal price diffusion with deterministic interest rate and dividend yield leads to the Black-Scholes pricing formula for equity options; and we need solve no PDE in the process. If the deterministic dividend yield is replaced by a set of discrete payments, or a richer parametrization of the dividend curve, then moment-matching methods can yield an inexact but quite accurate price.

As the models grow richer, the semianalytic methods must be correspondingly more sophisticated and complex. This is the "high art" of the quant world. In this volume, we will focus more on the frame: how can we make these approximations available without sacrificing genericity?

14.1 A Moment-Matching Pricer

Consider an equity process, so that the log price is normally distributed at some forward time. If the equity then pays a proportional dividend, the variance of this distribution is unaffected. However, a dividend of fixed size will tend to increase the variance. Suppose that the dividend amount is proportional to S^β for some $\beta \in [0, 1]$; a fixed dividend is simply the $\beta = 0$ limit. If we let $V(t)$ be the accumulated variance of the log at time t and Y the forward dividend yield, we find

$$\frac{\partial V}{\partial t} = \sigma^2 + 2(1 - \beta)VY,$$

plus higher-order terms (*e.g.*, from the higher moments of the true distribution at time t). To a similar approximation, the third moment of $\log S_t$ is governed by

$$\frac{\partial \xi}{\partial t} = \sigma^2 - 2(1-\beta)^2 V^2 Y.$$

So with these model dynamics, we can use this to price an equity option using only a sum over discrete dividends, or at worst a solution of two ordinary differential equations.[1] We will compute three moments of the distribution of the log; convert them to three moments of the distribution of the stock price at expiry; fit a shifted lognormal distribution to those as in Sec. 7.7.1; and price the option based on this distribution.

EquityOptionByMoments.cpp

```
    double OptionValue
        (const YieldCurve_& yc,
        double s_0,
        const Dividends_& divs,
5       const PWC_& vols,
        const Time_& expiry,
        long delivery,
        double strike,
        const OptionType_& type)
10  {

        const Vector_<> moments = MomentsOfExponent
                (AccumulateMoments(vols, divs, expiry, 3));
        assert(moments.size() >= 4);    // 0, 1, 2, 3
        double shift, shiftedFwd, shiftedVol;
15      FitThreeMomentsSLN
                (moments[1], moments[2], moments[3],
                 &shift, &shiftedFwd, &shiftedVol);
        return Distribution::Black_(shiftedFwd, shiftedVol, 1.0)
                .OptionPrice(strike - shift, type);
20  }
```

14.2   Multimethod Objects

How are we to reach this code from a generic function that receives only a `Trade_` and a `Model_`? Because we must check the type of both trade and model (and sometimes their internal properties as well), such a pricing request requires a *multimethod* – a function which is "virtual" in more

---

[1]This is a bad model in several ways - equity skew is not really explained by non-proportional dividends – but it illustrates our approach.

than one of its inputs. Little optimization is possible, because the internal switching process (however we control it) is inevitably irregular; there will be a patchwork of methods with very few sweeping generalities.

Thus we must prepare to query methods in sequence. For this purpose, we will add interface functionality to our trades and models, using mixins; then a given *candidate pricer* can inspect them with `dynamic_cast` to see whether they are suitable to its needs. All candidate pricers derive from

Semianalytic.h

```
struct SemianalyticPricer_
{
   virtual bool Attempt
      (_ENV, const Trade_& trade,
       const Model_& model,
       const Valuation::Parameters_& params,
       Vector_<pair<String_, double> >* vals)
   const = 0;
};
```

The derived classes should contain no member data – for which they can have no valid use – and the lack of a destructor is a reminder of this. For instance, a pricer to attempt the equity valuation above would look like

EquityOptionByMoments.cpp

```
// deprecated, fat interface
struct PriceByMoments_ : SemianalyticPricer_
{
   bool Attempt
      (_ENV, const Trade_& trade,
       const Model_& model,
       const Valuation::Parameters_& params,
       Vector_<pair<String_, double> >* vals)
   const
   {
      scoped_ptr<EquityOptionData_> opt
            (EquityOption::NewData(trade));
      DYN_PTR(myModel, const BlackWithDividends_, &model);
      if (!myModel || !opt.get())
         return false;
      const String_& eq = opt->eqName_;
      const double value = OptionValue
            (*model.YieldCurve(trade.valueCcy_),
             myModel->Spot(eq), myModel->Dividends(eq),
             myModel->Vols(eq), opt->expiry_,
             opt->delivery_, opt->strike_, opt->type_);
      vals->push_back(make_pair
```

```
                      (trade.valueNames_[0], value));
            return true;
25       }
    };
```

Here we have showed two ways the pricer can query its inputs: by direct **dynamic_cast**, or by calling a utility function which will (presumably) cast internally and return the relevant information upon success.

The resulting code is subpar, because we have made the model disclose too much of its data, and have used the resulting overly fat interface to create model-specific functionality located away from the model it depends on. Continuing along this path would require us to create similar pricers for each equity model, repeating our mistake.

We can remedy this by moving this version of **OptionValue** to the model's own source file, and rewriting the candidate pricer as

```
──────────── EquityOptionSemianalytic.cpp ────────────
struct PriceEquityOption_ : SemianalyticPricer_
{
   bool Attempt
      (_ENV, const Trade_& trade,
5       const Model_& model,
        const Valuation::Parameters_&,
        Vector_<pair<String_, double> >* vals)
   const
   {
10     scoped_ptr<EquityOptionData_> opt
            (EquityOption::NewData(trade));
       DYN_PTR(myModel, const HasAnalyticEquity_, &model);
       if (!myModel || !opt.get())
          return false;
15     vals->push_back(make_pair
            (trade.valueNames_[0], myModel->Price(*opt)));
       return true;
   }
};
```

This code will now be invoked for any model that (by deriving from **HasAnalyticEquity_**) declares that it can price an equity option. The **EquityOptionData_** must be declared where both models and trades can see it; our preference is to group it with the protocols described in Ch. 10.

14.3   Method Registry

Now we need a top-level function which can **Attempt** each candidate pricer until one succeeds. We store a run-time singleton registry of such functions, indexed by priority; thus the most common and fastest pricers can be checked first. We use **greater<int>** as the registry's sort criterion, so that large numbers mean high priority (we could also use reverse iterators for this).

─────────────── *Semianalytic.cpp* ───────────────
```
Vector_<pair<String_, double> > ValueSemianalytic
   (_ENV, const Trade_& trade,
     const Model_& model,
   const Valuation::Parameters_& params)
{
   const auto& candidates = ThePricers();
   Vector_<pair<String_, double> > retval;
   for (auto pc = candidates.begin();
        pc != candidates.end(); ++pc)
   {
      if (pc->second->Attempt
               (_env, trade, model, params, &retval))
      {
         return retval;
      }
      retval.clear();
   }
   Throw(_env, "Can't find any pricer");
}
```

We support registration (and, if necessary, deregistration) with a macro allowing users to write, *e.g.*,

─────────────── *EquityOptionSemianalytic.cpp* ───────────────
```
REGISTER_SEMIANALYTIC(PriceEquityOption_, 5)
```

14.4   Interaction with Re-evaluator

In Sec. 10.9.1, we showed a **ReEvaluator_** for stable pricing during bumped runs. Many semianalytic methods need no such stabilization, but any that can deviate from a single internal pricing path – *e.g.*, by using an adaptive integrator or ODE solver – will be subject to instability if the bumped run does not exactly follow in the footsteps of its base.

Rather than expand the interface of `ValueSemianalytic`, we pass an auditor through its environment, which will allow candidate pricers to store and later retrieve information.

```
———————————— Semianalytic.cpp ————————————
class ReevaluateSemianalytic_ : public ReEvaluator_
{   // member order used in initialization!
    AuditorImp_ auditor_;
    Environment::XEphemeral_ env_;
5   const Trade_& trade_;
    const Valuation::Parameters_& params_;

    ReevaluateSemianalytic_
        (_ENV, const Trade_& trade,
10        const Model_& model,
         const Valuation::Parameters_& params)
      : env_(_env, auditor_), trade_(trade), params_(params)
    {
        baseVals_ = ValueSemianalytic
15            (&env_, trade, model, params);
        auditor_.mode_ = AuditorImp_::SHOWING;
    }

    Vector_<pair<String_, double> > BumpedValues
20        (const Model_* bumped_model)
    {
        return bumped_model
            ? ValueSemianalytic
                  (&env_, trade_, *bumped_model, params_)
25            : baseVals_;
    }
};
```

Now any semianalytic method making a potentially discontinuous change in pricing – such as an integrator deciding how many points to use – can store and later recall its value using the techniques of Sec. 5.6.

14.5   Interaction with Composites

If the components of a composite trade can all be valued semianalytically, then we should be able to similarly value the composite trade itself. We accomplish this with a candidate pricer which calls back into `ValueSemianalytic`:

```
──────── SemianalyticComposite.cpp ────────
struct PriceComposite_ : SemianalyticPricer_
{
   bool Attempt
      (_ENV, const Trade_& trade,
       const Model_& model,
       const Valuation::Parameters_& params,
       Vector_<pair<String_, double> >* vals)
   const
   {
      DYN_PTR(composite, const CompositeTrade_, &trade);
      if (!composite)
         return false;
      auto subTrades = composite->SubTrades();
      for (auto ps = subTrades.begin();
              ps != subTrades.end(); ++ps)
      {
         Append(vals, ValueSemianalytic
               (_env, **ps, model, params));
      }
      *vals = composite->FinalValues(*vals);
      return true;
   }
};
```

A sharp-eyed reader may already have noticed the possibility of an unpleasant interaction between `ReevaluateSemianalytic_` and `PriceComposite_`: what if the same pricer code stores auditing information for more than one trade? To avoid this, we have to tag the auditing information with the name or (better) the address of the trade being priced; this in turn requires extra information in structures like `EquityOptionData_`. Alternatively, we can choose a "good-enough" solution where we tag the auditing information with some trade information related to the choice being remembered – e.g., a strike and/or expiry – which will almost certainly be unique in practice.

14.6 Pure Pricers

There are some models – such as equity implied vol surfaces, swaption cubes, and copulas – which do not support any numerical pricing; they simply postulate a distribution of some asset price, in the appropriate measure, at a future time when it will be needed for pricing. These *pure pricers* (also called *effective models*) fit very naturally into our framework: they derive from `Model_` but can never produce an `SDE_`. Thus their entire complexity is marshalled in support of some `Attempt` at direct semianalytic pricing.

14.7 Trade-Dependent Calibration

Our `SDE_` is customized to the trade being priced in a very weak way: it will avoid modeling underlyings which do not contribute to the trade, and may truncate yield curves when possible to save computation. We can envision a system of much stronger trade-dependence, where some detailed features of the trade are used to drive a *trade-dependent calibration* followed by numerical pricing in the calibrated model.

This is a distasteful practice: it reduces the internal consistency of our prices for different products, and distorts our aggregation of market risks across trades. However, if our models are too weak to fit the market with a single calibration, allowing a trade-dependent calibration can mask this weakness.

Models of this sort will produce a new `SDE_` for every trade request, whereas a simple model like VHW will always produce the same `SDE_`. This is why `ForTrade` should return a `Handle_`, which may be to a new or a cached object.

A common example is the USD Bermudan swaption market, which is liquid enough that we can accurately observe prices across a wide range of maturities, first-call tenors, and strikes. We might fit the European market with an effective model simply containing a cube of Black or SABR vol parameters; but we cannot price Bermudans without using an SDE.

A common solution is to calibrate a one-factor model to the *European components* of the Bermudan swaption – the options with the same terms as the Bermudan but restricted to a single expiry – or even to the single most valuable European component only.[2] We would use VHW (Sec. 13.1) for this purpose; thus the model, when asked to create an SDE for a trade, would calibrate a new VHW model and give its SDE.

To make this work, we need more information in `Underlying_`. We do not want to add any `Trade_` or `TradeData_` members – that would defeat our purpose of making `Underlying_` a low-level protocol by which a trade

---

[2]The latter approach will occasionally show unstable risk, whenever a market change or bump causes a different component to become the most valuable.

can express its needs. But we do need to embed polymorphic information.

Our solution is a narrow interface class, really just a base for `dynamic_cast` queries, which we will define inside `Underlying_`:

```
———————— Underlying.h ————————
class Underlying_
{
   class Parent_ : noncopyable
   {
   public:
      virtual ~Parent_();
   };
   Handle_<Parent_> parent_;
   // continues as before
};
```

Most trades will provide an `Underlying_` without a `parent_`, thus not volunteering any information which could drive a recalibration.

A Bermudan option itself cannot tell which component might be the most valuable (it should not even be able to tell which ones have expired), so it must disclose them all regardless of the recalibration being used. We support this with a new kind of `Parent_`:

```
———————— BermudanSwaption.h ————————
class HasEuropeanComponents_ : noncopyable
{
public:
   virtual Vector_<Handle_<Swaption_> >
        EuropeanComponents() const = 0;
};
```

The function to obtain the components will probably rely on `friend` access to the `BermudanSwaption_` class.

Once this is written, the pricer can obtain the options for recalibration. Suppose that we have a pure pricer class, `SwaptionCube_`, and are extending it to support Bermudan options in the manner described. Then we add the code

```
———————— SwaptionCube.cpp ————————
Handle_<SDE_> SwaptionCube_::ForTrade
   (_ENV, const Underlying_& underlying) const
{
   auto berm = HandleCast<HasEuropeanComponents_>
        (underlying.parent_);
   Require(0, !berm.Empty(),
        "Can't price non-Bermudan option numerically");
```

```
       auto euros = berm->EuropeanComponents();
       assert(!euros.empty());
10     auto euroVals = Apply(XSwaptionPrice_(*this), euros);
       const int iMVE = MVEIndex(_env, euroVals);
       const Swaption_& mve = *euros[iMVE];
       scoped_ptr<VHW_> vhw
              (VHW::MatchSwaptionHoLee
15                    (YieldCurve(mve.valueCcy_), mve,
                       euroVals[iMVE]));
       return vhw->ForTrade(_env, underlying);
}
```

This relies on a factory function **NewHoLee** which calibrates a Ho-Lee model's single volatility parameter to one European option. The **SwaptionCube_** must support some candidate pricer which can be called by **ValueSemianalytic**; the supporting class **SwaptionPrice_** can call this pricing function directly, rather than meander through **ValueSemianalytic**'s search over candidates.

14.7.1 *Stabilization*

The helper function **MVEIndex** can simply return the index of the maximum element; but this will lead to occasional explosions in computed risk figures, whenever a small bump happens to cause a different component to become the most valuable. Our stabilization mechanism can prevent this:

SwaptionCube.cpp

```
int MVEIndex(_ENV, const Vector_<>& euro_vals)
{
    static const String_ KEY("MVEIndexForBerm");
    Handle_<SaveMVEIndex_> store;
5   Environment::Recall(_env, KEY, &store);
    if (store.Empty())
    {
        int i = MaxElement(euro_vals) - euro_vals.begin();
        store.reset(new SaveMVEIndex_(i));
10      Environment::Audit(_env, KEY, store);
    }
    return store->i_;
}
```

Here **SaveMVEIndex_** is a **Storable_ struct** whose sole data member is the index **i_**.

Chapter 15

Risk

To manage a derivatives book, we must measure its exposure to market prices and to less observable parameters – its *risks* – systematically and accurately. If we are to retain flexibility of models, we must ensure that risk is not mapped to model-specific parameters; thus we emphasize the creation of uniform specifications of a change to a model, and think of risk computation as mostly a process of repeated valuation with a series of slightly different models.

15.1 Slides and Bumps

We must specify a change while attempting to avoid specifying the structure of the model to which it will be applied. Such changes are usually described as *slides* – large changes to measure our exposure to a substantial market move – or small *bumps* to measure a first-order sensitivity to a hedge parameter.

In the context of their application, these are indeed different: for instance, the stabilization mechanisms of Secs. 10.0.1 and 14.4 should always be applied for bumps – so that valuation differences will be only those caused directly by model parameter changes and never for slides, where the stabilization mechanisms will inevitably suppress the increasingly important higher-order terms. But to a model, this difference is invisible; here we use "slide" to describe a bump of any type or size.

Thus a slide is, in principle, a function which takes one (fully parametrized) input model, and produces a different model. But an implementation such as

```
// deprecated
```

```
class Slide_ : public Storable_
{
public:
    virtual Handle_<Model_> operator()
        (const Handle_<Model_>& src)
    const = 0;
};
```

will quickly become unworkable, as it forces slides to cope directly with the dynamics of every model we might use. We must instead have the low-level `Slide_` object display its data, and the high-level `Model_` decide how to respond.

For this purpose, it is crucial that *slides must be simple atoms* – in particular, they cannot be decorated or composite objects. If a model determines (probably with `dynamic_cast`) that a given slide affects, *e.g.*, a yield curve, then it should be allowed to assume that the same slide does not also request the manipulation of equity vols. Slides will indeed be composed, but such a composite (which we call a *scenario*) will be represented as a `vector` of individual, atomic slides.

15.2   Mutants

Applying a slide obviously must change the model's contents. We do not want a non-`const` `Apply` function, which would mutate a model in-place; instead, we construct the new `Model_` as a new object. This is the role of `Mutant_Model` from Sec. 10.13.4: The interface (in class `Model_`) is

```
———————————— Model.h ————————————
    virtual Model_* Mutant_Model
        (const String_& new_name,
         const Vector_<Handle_<Slide_> >& slides)
    const = 0;
```

The function is called `Mutant_Model`, not simply `Mutant`, because of the confusion sown by many unrelated methods all named `Mutant`; also, we will often support its implementation in concrete classes with further mutators, which we can then name `Mutant_VHW` etc.

We pass several slides at once as a potential optimization, since this might prevent the full construction of intermediate models. It is easy enough for a derived class to implement this in terms of a simpler function, eschewing the optimization:

```
─────────────── VHWModel.cpp ───────────────
Model_* VHW_::Mutant_Model
    (const String_& new_name,
    const Vector_<Handle_<Slide_> >& slides)
const
{
    auto_ptr<VHW_> retval(Mutant_VHW(&new_name, 0));
    for (auto p = slides.begin(); p != slides.end(); ++p)
    {
        auto_ptr<VHW_> t(retval->Mutant_VHW(0, p->get()));
        swap(retval, t);
    }
    return retval.release();
}
```

Note that this implementation assumes slides are applied front-to-back, as users will invariably expect.

15.3 Reports

In a given computation, we will extract sensitivities of some set of values (because of composite trades, we must always consider the possibility of multiple values), to some set of parameters. The parameters may be organized along more than one axis – for example, a local vol sensitivity would be indexed by time and by strike. Finally, we may choose to report multiple *views* of the same data as a service to our users; for instance, we might report both a raw sensitivity and the notional amount of an instrument required to hedge it. Thus a risk report must organize information along multiple axes – up to at least five.

Since the report lacks a fixed dimension, we might as well store its entries in a single container; we use **deque** rather than **vector** to avoid demanding a vast amount of contiguous memory.[1] Thus we begin our implementation with

```
─────────────── Report.h ───────────────
class Report_ : public Storable_
{
    map<String_, int> axes_;     // lookup location
    Vector_<int> strides_;       // back() is whole size
    deque<double> vals_;
public:
```

─────────────────────────

[1] A sparse storage scheme might be better for the largest reports; we do not display one here.

```
     Vector_<String_> Axes() const {return Keys(axes_);}
     int Size(const String_& axis) const;

10   typedef Report::Address_ Address_;
     double& operator[](const Address_& loc);
     const double& operator[](const Address_& loc) const;
     // ...
```

The layout is elucidated by a look at `Size`:

```
─────────────── Report.cpp ───────────────
int Report_::Size(const String_& axis) const
{
    REQUIRE0(axes_.count(axis), "No axis '" + axis + "'");
    const int which = axes_.find(axis)->second;
5   return strides_[which + 1] / strides_[which];
}
```

We translate the axis name to an index using the lookup `axes_`; later axes walk through `vals_` with successively larger `strides_`.

We encapsulate element access in a single-argument `operator[]` which accesses a single `Address_` value; there are two viable implementations. The easiest way is to define an address which is essentially a `map<String_, int>`; this supports idioms like

```
for (loc[TRADE] = 0; loc[TRADE] < nTrades; ++loc[TRADE])
```

where `loc` is the address. At each call to the report's `operator[]` (not the address's), we will translate the string keys to corresponding axes.

It is slightly more efficient to have the address type store the axis ordering; then the lookup occurs only in the address's `operator[]`, which is called less often. The price is that we cannot construct such an address until we have an instance of the report.

```
─────────────── Report.h ───────────────
namespace Report
{
    struct Address_
    {
5       map<String_, int> axes_;
        vector<int> locs_;
        Address_(const map<String_, int>& axes)
            : axes_(axes), locs_(axes.size()) {}
        int& operator[](const String_& axis);
10  };
```

```
   }

   class Report_ : public Storable_
   {
15     // ...
       Address_ MakeAddress() const;
       double& operator[](const Address_& loc)
       {
           assert(loc.locs_.size() == strides_.size() - 1);
20         return vals_[inner_product(loc.locs_.begin(),
                   loc.locs_.end(), strides_.begin(), 0)];
       }
       // ...
```

15.3.1 *Barewords*

We have used **TRADE** as a variable name above; this is actually the name of a constant defined by

──────────── *ReportUtils.h* ────────────

```
namespace ReportAxes
{
    BAREWORD(TRADE);
    BAREWORD(VIEW);
5   // ...
}
using namespace ReportAxes;
```

This makes our meaning clear while enlisting the compiler to check for misspellings.

15.4 Portfolios

So far we have talked about individual trades; also, we have discussed only the financial content of trades, ignoring bookkeeping issues such as the need to record the trader and counterparty. Also, we have talked about valuation parameters without describing their origin.

We will use *portfolio* to mean a collection of trades, plus additionally for each trade:

- Bookkeeping information, which we may need to reflect in a risk report.
- Valuation parameters controlling the running of risk.

A portfolio, as stored by a system, will likely associate further information with each trade, such as a rule for finding the model with which to value it.

Since bookkeeping information must be communicated from a Portfolio_ to a Report_, we add to Report_ the members

```
——————————————— Report.h ———————————————
    Vector_<pair<String_,Vector_<String_> > > bookkeeping_;
public:
    void AddBookkeeping(const String_& title,
        const Vector_<String_>& values)
5   {
        assert(values.size() == Size(ReportAxes::TRADE));
        bookkeeping_.push_back(make_pair(title, values));
    }
```

which the Portfolio_ can write to. Thus we need

```
——————————————— Portfolio.h ———————————————
class Portfolio_ : public Storable_
{
    Vector_<Handle_<Trade_> > trades_;
    Vector_<Valuation::Parameters_> params_;
5   Vector_<TradeBookkeeping_> books_;
public:
    const Trade_& Trade(int i_trade) const;
    const Valuation::Parameters_& ValuationParams
        (int i_trade) const;
10  void WriteBookkeeping(Report_* dst) const;
};
```

15.5 Tasks

Having described what must be the output of a risk run, we examine its inputs. It must take a portfolio and model; in the process of pricing, these may make some demands on the environment (*e.g.*, for the presence of certain fixings).

Thus we define a generic risk task

```
——————————————— Risk.h ———————————————
class RiskRun_ : public Storable_
{
public:
    virtual Report_* Run
5       (const Portfolio_& portfolio,
         const Model_& model)
```

```
    const = 0;
};
```

This class must be **Storable_** because a call to **Run** is the atom of a distributed computation.[2]

15.6   Slide Utilities

Many objects will be shared across model types; for slides which can be implemented as changes to these objects, we place this implementation in a utility function for models to call as needed.[3] The canonical example is the yield curve, shared by essentially all models. For this we provide a shared utility

```
──────── SlideIR.h ────────
namespace SlideIr
{
    bool Apply
        (Vector_<Handle_<YieldCurve_> >* curves,
5       const Slide_& slide);
}
```

The subtlety here is that there are two ways in which the slide can have no effect on curves: if it is not an interest-rate slide, or if it slides some curve not present in the model. In the former case, the model must continue inspecting the slide to see if it affects some other parameter; we can signal this by returning **false** from **Apply**. In the latter case, we are done and the slide has no effect.

Thus the calling code will look like

```
    if (SlideIr::Apply(&newCurves, slide))
    {
        changed = changed || YieldCurves() != newCurves;
        // act on changed curves
5   }
    else
        // check some other kind of slide
```

---

[2]We can sometimes split the **Run** itself, but generic techniques for this are beyond this volume's scope.

[3]And *never* in base classes for derived classes to call.

Since slides will be accumulated to effect a scenario, this code will likely be inside a loop over slides, and if `Apply` succeeds we will simply `continue`.

15.7   Conclusions

Slides provide us with a general way to perturb any model that can respond to them; thus we can construct risk computations without direct reference to the model type. The techniques of Ch. 10 let our trades communicate with *any* model in numerical pricing, allowing us to prototype new models on existing trades and perform alternative risk analyses. Ch. 14 shows how we can insert fast semianalytic methods wherever they are appropriate, and preserve the stability of risk.

We have worked carefully to separate trades and models; to enable component-based models for prototyping and hybrids; and to provide flexible methods for preserving stability of risk regardless of the details of pricing. These are crucial steps on the road to analytics superiority.

Chapter 16

Additional Code

16.1 Add Multiple

```
                          Numerics.h
   template<class T_> struct AddMultiple_ :
        binary_function<T_, T_, T_>
   {
      T_ a_;
5     AddMultiple_(const T_& a) : a_(a) {}
      T_ operator()(const T_& x1, const T_& x2) const
      {
          return x1 + a_ * x2;
      }
10 };
   template<class T_>
   AddMultiple_<T_> AddMultiple(const T_& a)
   {
      return AddMultiple_<T_>(a);
15 }
```

16.2 ArrayFunctor

```
                          Vectors.h
   template<class C_, class R_> struct XArrayFunctor_ :
        unary_function<int, typename C_::value_type>
   {
      R_ val_;
5     XArrayFunctor_(const C_& val) : val_(val) {}
      typename const C_::value_type& operator()(int ii)
          const {return val_[ii];}
   };

10 template<class C_> XArrayFunctor_<C_, const C_&> XLookupIn
```

291

```
      (const C_& src)
   {
      return XArrayFunctor_<C_, const C_&>(src);
   }
15 template<class C_> XArrayFunctor_<C_, C_> LookupIn
      (const C_& src)
   {
      return XArrayFunctor_<C_, C_>(src);
   }
```

Note that **XLookupIn** returns an ephemeral structure, with a stored reference to the input array.

16.3 Boolean

```
――――――――――― Boolean.h ―――――――――――
template<class T_> struct Boolean_
{
   T_ val_;
   Boolean_(const T_& val) : val_(val) {}
5  T_& operator=(const T_& rhs) {return val_ = rhs;}
   operator bool() const {return TruthValueOf(val_);}
   operator const T_&() const {return val_;}
};
```

16.4 Composite

```
――――――――――― Composite.h ―――――――――――
template<class T_> class Composite_ : public T_
{
   vector<shared_ptr<T_> > contents_;
public:
5  void Append(T_* p)
   { contents_.push_back(shared_ptr<T_>(p)); }
   int Size() const {return contents_.size();}
   int Empty() const {return contents_.empty();}

10   T_* operator[](int i)
   {
      assert(i < contents_.size());
      return contents_[i].get();
   }
15   const T_* operator[](int i) const
   {
```

```
        assert(i < contents_.size());
        return contents_[i].get();
    }

    // Default constructible
    Composite_() {}
    // Copy constructible if T_ is clonable
    Composite_(const Composite_<T_>& src)
    {
        for (int ic = 0; ic < src.Size(); ++ic)
        {
            Append(src.contents_[ic]
                   ? src.contents_[ic]->Clone()
                   : 0);
        }
    }
    Composite_<T_>* Clone() const
        {   return new Composite_<T_>(*this);   }
    // Assignable if T_ is
    T_& operator=(const T_& rhs);
},

template<class T_> Composite_<T_>& CastComposite(T_* src)
{
    return dynamic_cast<Composite_<T_>*>(src);
}
template<class T_> Composite_<T_>& CoerceComposite(T_* src)
{
    assert(dynamic_cast<Composite_<T_>*>(src));
    return static_cast<Composite_<T_>&>(*src);
}
// ... also const version
```

16.5 Cube

See Sec. 7.9.1.

16.6 Handle

```
────────────── SmartPtr.h ──────────────
template<class T_> class Handle_ :
    public shared_ptr<const T_>
{
```

```
   public:
 5    Handle_() {}
      explicit Handle_(const T_* src)
            : shared_ptr<const T_>(src) {}
      Handle_(const shared_ptr<const T_>& src)
            : shared_ptr<const T_>(src) {}
10    Handle_(const shared_ptr<T_>& src)
            : shared_ptr<const T_>
                    (shared_dynamic_cast<const T_>(src)) {}

      bool Empty() const {return !get();}
15 };
   template<typename T_> bool TruthValueOf
      (const Handle_<T_>& src) {return !!src.get();}

   template<class T_, class U_> Handle_<T_> HandleCast
20    (const shared_ptr<const U_>& src)
   { return shared_dynamic_cast<const T_>(src); }
```

16.7   Matrix

───────────── *Matrix.h* ─────────────
```
   template<class E_ = double> class Matrix_
   {
      typedef typename vector<E_>::iterator I_;
      vector<E_> vals_;
 5    vector<I_> hooks_;
      int nColumns_;
      void SetHooks(int from = 0)    // see Resize()
      {
         for (int i = from; i < hooks_.size(); ++i)
10          hooks_[i] = vals_.begin() + i * nColumns_;
      }
   public:
      virtual ~Matrix_() {}
      Matrix_() : nColumns_(0) {}
15    Matrix_(int rows, int cols) :
          vals_(rows * cols), hooks_(rows), nColumns_(cols)
      { SetHooks(); }
      Matrix_(const Matrix_<E_>& src);    // calls SetHooks
      Matrix_<E_>& operator=(const Matrix_<E_>& src);
20    void Swap(Matrix_<E_>* other);
      inline void Fill(const E_& val); // calls std::fill
      void Resize(int rows, int columns);
      // Properties
```

```
     int Rows() const {return hooks_.size();}
25   int Columns() const {return nColumns_;}
     bool Empty() const {return hooks_.empty();}
     typename vector<E_>::const_iterator Last() const
         {return vals_.end();}    // to detect aliasing
     // Fortran-style addressing for maximum speed
30   typename vector<E_>::const_reference operator()
         (int row, int col) const {return hooks_[row][col];}
     typename vector<E_>::reference operator()
         (int row, int col) {return hooks_[row][col];}
     // Slices -- so rows look like independent containers
35   class ConstRow_
     {
     protected:
         // non-const to support Row_, below
         I_ begin_, end_;
40   public:
         typedef E_ value_type;
         typedef I_ iterator;
         typedef typename vector<E_>::const_iterator
             const_iterator;
45       // construct from begin/end or from begin/size
         ConstRow_(I_ begin, I_ end)
             : begin_(begin), end_(end) {}
         ConstRow_(I_ begin, int size)
             : begin_(begin), end_(begin + size) {}

50
         const_iterator begin() const {return begin_;}
         const_iterator end() const {return end_;}
         int size() const {return end_ - begin_;}
         const E_& operator[](int col) const
55               {return *(begin_ + col);}
     };
     ConstRow_ Row(int row_index) const
         {return ConstRow_(hooks_[row_index], nColumns_);}
     ConstRow_ operator[](int row_index) const
60       {return Row(row_index);}    // matrix-style access

     struct Row_ : public ConstRow_
     {
         // Inherit data members and const_iterator type
65       // Implement constructors and non-const iterator
         ...
     };
```

```
      // Iterate through columns as well (but less efficient)
70    class ConstColumn_
      {
      protected:
          // non-const to support Column_, below
          I_ begin_, end_;
75    public:
          typedef E_ value_type;
          typedef I_ iterator;
          typedef typename vector<E_>::const_iterator
              const_iterator;
80        // construct from begin/end or from begin/size
          ...
      };
      ConstColumn_ column(int column_index) const ...
      class Column_ : public ConstColumn_    ...
85
      // A submatrix sees our data, or part thereof
      class SubMatrix_
      {
          vector<T_> hooks_;
90        const int nColumns_;
      public:
          int Rows() const {return hooks_.size();}
          int Columns() const {return nColumns_;}
          bool Empty() const {return hooks_.empty(); }
95
          // Fortran-style addressing for maximum speed
          typename vector<E_>::const_reference operator()
              (int row, int col) const
          { return hooks_[row][col]; }
100       typename vector<E_>::reference operator()
              (int row, int col) {return hooks_[row][col];}

          ConstColumn_ Column(int ic) const
              {return Column_(hooks_[ic], nColumns_);}
105       Column_ Column(int ic)
              {return Column_(hooks_[ic], nColumns_);}
          // ...
      };

110   // Return an "ephemeral" submatrix
      // (becomes invalid when *this is destroyed or resized)
      SubMatrix_ SubMatrix
          (const pair<int, int>& row_range,
```

```
                 const pair<int, int>& column_range);
115      const SubMatrix_ SubMatrix
             (const pair<int, int>& row_range,
                 const pair<int, int>& column_range) const;
         explicit Matrix_(const SubMatrix_& src) ...
     };
```

It seems that it should be possible to have **Matrix_** inherit from **SubMatrix_**, adding the resizing functionality. Then most functions taking a **const Matrix_** input could be changed to take a **const SubMatrix_** instead; perhaps a more euphonious name than **SubMatrix_** would then be preferable. We have not implemented this in a fully acceptable way.

16.8 Maybe

This is the OCaml/F# option type.

—————— *Maybe.h* ——————

```
template<class T_> class Maybe_
{
    T_ val_;
    bool known_;
5 public:
    typedef T_ value_type;
    Maybe_(const T_* val = 0) : known_(false)
        { if (val) Set(*val); }
    Maybe_(const T_& val) : known_(true), val_(val) {}
10  void Set(const T_& val) {val_ = val; known_ = true;}
    void Clear() {known_ = false;}
    bool Known() const {return known_;}
    const T_& Value() const {assert(known_);return val_;}
    const T_& Value(const T_& v0) const
15      { return known_ ? val_ : v0; }
    const T_* Pointer() const {return known_ ? &val_ :0;}
};
template<class T_> T_ operator+(const Maybe_<T_>& rhs)
{
20  return rhs.Known() ? rhs.Value() : T_();
}
template<typename T_> bool TruthValueOf
    (const Maybe_<T_>& src) {return src.Known();}
```

The overloaded unary + operator is a valuable shorthand, *e.g.* for optional data members; it supplies a default value by default construction. There is no implicit conversion of `Maybe_` to `bool` – it is too dangerous. Instead we implement `TruthValueOf` – see Sec. 16.3.

16.9 PWC (Piecewise Constant)

PWC.h

```
class PWC_ : public Storable_
{
public:
    struct Piece_
    {
        double f_;
        double cumF_;
    };
private:
    map<Time_, Piece_> vals_;
public:
    PWC_
        (const Vector_<Time_>& t,
         const Vector_<double>& f,
         const String_& name = String_());

    const map<Time_, Piece_>& Vals() const {return vals_;}
    operator const map<Time_, Piece_>&() const
    { return Vals(); }

    void Save(Archive::Dst_&) const {}
};
```

This is made more useful by some utility functions which query a `PWC_` object.

PWCUtils.h

```
namespace PWC
{
    double F(const PWC_& func,
        const Time_& t,
        double* is_knot = 0);
    double IntegrateF(const PWC_& func,
        const Time_& from,
        const Time_& to);
    PWC_ Constant
        (double val,
```

```
                const String_& name = String_(),
                const Time_& from = Time::Minimum());
    }
```

Efficiency of **IntegrateF**, which is very heavily used, is why we store the cumulant **cumF_** in our class.

16.10   Vector

```
                         ─── Vectors.h ───
   // we do not need all the template parameters of vector
   template <typename E_ = double> class Vector_ :
         private vector<E_>
   {
 5 public:
      Vector_() {}
      explicit Vector_(int size, const E_& fill = E_());
      template<class I> Vector_(I start, I stop)
          : vector<E_>(start, stop) {}
10    Vector_(const vector<E_>& src) : vector<E_>(src) {}
      using vector<E_>::assign;
   // ...
      void resize(int new_size)
          { this->vector<E_>::resize(new_size); }
15    // two arg resize, which does NOT fill, not provided
      // Added functionality
      Vector_& operator+=(const Vector_<E_>& increment);
      Vector_& operator-=(const Vector_<E_>& increment);
      Vector_& operator*=(const E_& scale);
20    // inefficient /= is not provided
      Vector_& operator+=(const E_& shift);
      Vector_& operator-=(const E_& shift);
      void Fill(const E_& v) {std::fill(begin(), end(), v);}
      void Append(const Vector_<E_>& v)
25       { insert(end(), v.begin(), v.end()); }
   };
```

300 *This page intentionally left blank*

Acknowledgements and Further Reading

The views and opinions expressed in this book are those of the author and are not those of UBS AG, its subsidiaries or affiliate companies ("UBS"). No relationship, association, sponsorship or endorsement is suggested or implied between UBS and the author through the publication of distribution of this book. Accordingly, UBS does not accept any liability over the contents herein or any claims, losses or damages arising from the use of or reliance on all or any part of this book.

The standard text on physical code structure is John Lakos's *Large-Scale C++ Software Design*. The principles it sets down are echoed throughout this work.

The uses and ramifications of automatic interface generation were largely worked out by Ian Taylor at Bankers Trust in 1996-97.

Advanced techniques for numerical and semianalytic pricing are derived in Alex Lipton's *Mathematical Methods for Foreign Exchange*.

The routines of *Numerical Recipes*, by Press *et al.*, are a good place to start for many implementation problems. This is also a book to be explored, to learn what methods might be available for problems one has not yet met.

The challenge and promise of higher-level programming are often explored by advocates of non-C languages. I have particularly profited from Paul Graham's *On Lisp*; the *Objective Caml Manual* of Xavier Leroy *et al.*; Mark-Jason Dominus's *Higher-Order Perl* and the Perl 6 language specification; and Andrei Alexandrescu's *Modern C++ Design*.

I learned the importance of the separation of asset and payout from Neil Smout and Trevor Chilton at UBS, who laid the conceptual foundations of Ch. 10.

On a more personal level, I would like to thank many colleagues and friends, who have contributed in more diffuse but no less real ways; especially Leif Andersen, Jesper Andreasen, Christophe Chazot, Steve Dugdale, Andy Felce, Richard Gladwin, Don Goldman, Stewart Inglis, Sandeep Jain, Alan LeGuen, Alex Lipton, Chris Mitchell, Vladimir Piterbarg, Dmitry Pugachevsky, Gerson Riddy, Paul Romanelli, Gleb Sandmann, and Richard Waddington.

Index

30E/360 ISDA, 144

accounting date, 88, 177
AddMultiple, 291
algorithms
    container-level, 38
        Unique, 39
aliasing, 43, 46
American Monte Carlo, 134, *see also*
    backward induction
    biases, 136
    observables, 134
archive, *see also* storable
    build record, 64, 67
    builders, 68
    data, 63
    extraction, 62
    readers, 66
    repeated objects, 61
    splat, 75
ArrayFunctor, 291
asset, 176, 252
auditing, 82

backward induction, 176
    actions, 192
    fees, 192
bag, 82
BAREWORD, 72, 287
baskets, 115
Black-Karasinski model, 263
BLITZ++, 43

Boolean, 292
Boost, 3
Brent's method, *see* rootfinders
bumped model, 119
bumped models, 201

caps, 226
causality, 202
CDS, 242
CIR model, 262
code generation, 16, 17, 21
    enumerations, 33, 139
    for archive, 70
    public function, 14
compilation
    dependencies, 6, 24
composite trades, *see* trades
conjugate gradient, 52
conventions
    code style, 3
    financial, 7, 151, 160, 239
        holidays, 147
cube, 128
    storage, 128

date increments, 154
dates
    past and future, 177
decorator, 9
    uses, 29, 191, 200, 205, 232
default
    for enumeration, 35

for settings, 110
for system, 88
for template, 37
function arguments, 100, 113, 162
default constructor, 86, 194, 227, 298
    lack of, 83, 149, 153, 159
defaults, *see also* conventions
defaults (of credit), 23, 162, 176, 178,
    235, 242
    event on path, 185
discounting, 238
    funding adjustment, 239
distributions
    interface, 113
    robust implied vol, 115
dividend curves, 241
Dupire formula, 265, 266

enumerations, 33
    extensible, 139, 155
environment, 23, 82
    facts, 6
    in exceptions, 25
    macros, 25
    nested, 29
    platform, *see* platform
    repository access, 31
ephemeral classes, 4, 29, 41, 53, 297
Erlang, 59
exceptions, 25

factory method, 9
    uses, 97, 120, 211, 249
fixings, *see* indices
functional programming, 5
    activation, 5
future, *see* past

Handle, 293
hazard curves, 242
header files, 7
    platform, *see* platform
holidays, 147
Hull-White model, *see* Vasicek model
hybrid model, 268

indices, 165
    composite, 171
    fixings, 165, 168
    implied vol, 172
    names, 165
    parsing, 165
    paths, 176, 183, 187
    short names, 167
    sorting, 171
inline, 123
instrumentation, 94
interface, *see also* code generation
    public, 13, 15, 16
        types, 18
        validation, 21
interpolation
    cubic spline, 97

Java, 17
join, 40

libor curves, 238
    instruments, 239
local vol model, 265

mark-up, *see* code generation
matrices, 41
    decompositions, 45, 46, 48, 57
    multiplication, 43
    sparse, *see* sparse matrices
    storage, 41
Matrix, 41, 294
Maybe, 297
merge
    for matrix, 41
mixins, 46, 52, 210, 275
model
    asset, *see* asset
    components, 268
    stepper, *see* stepper
models
    bumped, 283
    effective, *see* pricers
    mutating, 284
Monte Carlo, 201
multimethods, 274

registry, 277

normal distribution, 99
  inverse, 99
    polishing, 100
NOTICE, 30
numerical integration, *see* quadrature
Numerical Recipes, 45

operators
  conversion, 44
optimization, 5
overloading
  function, 38
  operator, 43, 68

past
  definition of, 177
paths record, 196
pattern matching, 90
payment tag, 179
payments
  conditions, 180
  from payout, 182
  on default, 187
payout, 176, 194
  state, 176, 195
PDE solvers, 127, 199
  coefficients, 131
  coordinates, 129
  forward induction, 133
persistence, *see* storable
platform, 7
platform header, 8
portfolios, 287
preconditioning, *see* conjugate
  gradient
pricers, 273
  candidate, 275
  pure, 279
probability distributions, 113
promises, 190, *see* value request
protocols, 175
  class hierarchy, 203
public type set, 64
Python, 59

Q-form, 48
quadratic programming, 105
  overshoot and backtrack, 109
  QP step, 105
quadrature, 110
  adaptive, 113
  Gaussian, 112

RAII, 88
random numbers, 118
  correlated, 47
  low-discrepancy sequences, 123
    allocation, 126
  normal deviates, 121
  uniform deviates, 120
re-evaluator, 199, 277
repository, 87
RepositoryErase, 34
Require, 26
reset time, 177
risk
  curve delta, 237
  report object, 285
  slides, 289
  tasks, 288
root search, 101
  underdetermined, 104
rootfinders
  Brent's method, 102
  initial bracketing, 103
  interface, 102
  underdetermined
    backtracking, 109
    Jacobian, 106
    weights, 108

scenario, 290
scripting, 5, 87, 95, 210
semianalytic pricers, *see* pricers
singleton, 9
  uses, 15, 66, 68, 91, 95, 141, 148,
    154, 167
singletons, 145
slides
  atomic, 284
SOAP, 17

Sobol sequences, 124, *see also* random numbers
  dimension, 127
  direction numbers, 125
sparse matrices, 48
  row-indexed, 51
steppers, 176, 196
storable, 59, *see also* archive
  build record, 67
storable, 60
streams, 181
strings, 22
swap, 4
swap legs, 157
swaps, 222, 228
swaptions, 228
  Bermudan, 230

template algorithms, 38
template method
  labels, 4
  uses, 46
templates
  specialization, 42
testing, 91
  component, 91
  regression, 94
threads, 11, 196
Thursday bugs, 157
token
  of update, 189
trades
  base class, 209
  cash, 211
  composite, 231, 278
  equity, 215
  fx, 218
  payout, *see* payout

underlying, 178

value date, 177
value request, 176, 190, 252
  promises, 176
variant, 19, 26
Vasicek model, 245

numerical pricing, 249
Vector, 37, 299
visitor, 19
vol start time, 177